District Heating and Cooling Networks in the European Union

Antonio Colmenar-Santos
David Borge-Díez · Enrique Rosales-Asensio

District Heating and Cooling Networks in the European Union

Springer

Antonio Colmenar-Santos
Departamento de Ingeniería Eléctrica, Electrónica y Control
Universidad Nacional de Educación a Distancia (UNED)
Madrid
Spain

Enrique Rosales-Asensio
Departamento de Física
Universidad de La Laguna
San Cristóbal de La Laguna
Spain

David Borge-Díez
Departamento de Ingeniería Eléctrica y de Sistemas y Automática
Universidad de León
León
Spain

ISBN 978-3-319-57951-1 ISBN 978-3-319-57952-8 (eBook)
DOI 10.1007/978-3-319-57952-8

Library of Congress Control Number: 2017939606

Printed on acid-free paper

This Springer imprint is published by Springer Nature
The registered company is Springer International Publishing AG
The registered company address is: Gewerbestrasse 11, 6330 Cham, Switzerland

Acknowledgements

We are grateful to the experts contacted who gave us their support, in particular, and to Poul Alberg Østergaard, Capezzali Massimiliano, Kevin Sartor, Erwin Cornelis, Chiara Wolter, Alexandra Tudoroiu, Olu Ogunbadejo, Aleksandrs Zajacs, Carsten Magass, Helen Andrews Tipper, Daniel Wright, Robert Purdon, Peter Connor, Anthony Granville, David Hawkey, Torben Hermansen, Rasmus Aaen, Dag Henning, Olof Hjelm, David Appeblad, Monika Kuusela, Kohopää Antti, Anna Krook-Riekkola, David Andrews, Francisco José Macías Benigno, Luca Angelino, Ana Stojilovska, Andrey Viktorovich Kovalev, Meegan Kelly, Gunnar Boye Olesen, Poul Erik Grohnheit, Henrik Klinge Jacobsen, Erik Ahlgren, Mark Barrett, Ginés de Rus-Mendoza, Massimo Florio, Marco Ponti, Arturo Lorenzoni, Jonathan Healy, Aviel Verbruggen, Yamina Saheb and Marta Santamaría. Without their participation, this book would not have been possible.

Contents

Symbols, Abbreviations and Acronyms

Ac_t	Annualised capital costs (those costs incurred when constructing the power plant)
AHDD	Annual heat demand per dwelling
A(n, d)	Annuity factor for an equipment of *n* years and a discount rate *d*
AVAIL	Yearly availability
CBA	Cost–benefit analysis
c_t	Cost of capital for a given technology
d	Discount rate
DCNs	District cooling networks
DHNs	District heating networks
EC	European Commission
ECB	European Central Bank
f	Combustible expenses
HHW	Heating and hot water
HPR	Heat-to-power ratio
IRR	Internal rate of return
LAs	Local authorities
LF	Load factor
LS	Lifespan, time during which a continuous renewal of the facility is assumed
ms	Mass flow of steam
$m \cdot X$	Fixed running and preservation expenses (includes plant operations personnel, taxes, insurances, and other services)
n	Lifetime of technology
NPV	Net present value
PFI	Private finance initiative
r	Discount rate used to discount future cash flows to their present value
REV	Revenue resulting from the sale of power
R_t	Net cash flow (cash amount, input minus output) for a time *t*

SDR	Social discount rate
t	Time (in years) of cash flow relative to the start of the project
$v \cdot X$	Variable running and preservation expenses (includes variable maintenance and material costs)
X	Rated power of a given technology
η_{BP}	Ratio of the power generated and the heating value of the fuel consumed in backpressure mode mode
η_C	Refrigeration performance
η_e	Ratio of the power generated and the heating value of the fuel consumed in electricity-only mode
η_h	Thermal performance/network per heat added in back

List of Figures

List of Tables

Abstract and Aims

The purpose of this book was to investigate the possible potential of the joint use of district heating networks and cogeneration in the EU-28, propose measures to remove barriers hindering its widespread implementation, formulate policies for such implementation, and evaluate the economic, energy, and environmental consequences resulting therefrom.

To this end, a preliminary assessment of the likely cost and the impact of adopting the widespread use of district heating networks and cogeneration were carried out in three cities with different representative climatic conditions of the EU-28. From the results of this assessment, it could be estimated that for all the EU-28 and undertaking the maximum economically feasible implementation, a fuel savings of 95 milliard euros a year and a savings of the required primary energy of 6400 PJ would occur, all through an annual investment of 315 milliard euros. This represents about 15% of the total final energy consumption in the EU-28 in 2013 (46,214.5 PJ). This amount would be even greater if this book would have considered, apart from conventional thermal plants, nuclear power plants existing in the EU-28.

To achieve their full development, companies in charge of exploiting those schemes which jointly use district heating networks and cogeneration should enjoy quasi-governmental powers to create significantly reduced market risks and, therefore, reduced funding costs. It is essential that all economic evaluations of the joint use of district heating networks and cogeneration are carried out using an appropriate discount rate for infrastructural projects ("social" discount rate), which for projects that are located in Western Europe is 3.5% following the recommendations of the European Commission.

About 80% of the energy used in homes has as the ultimate goal of providing heat and hot water; this means that there would be potentially 15,000 PJ/year (4100 TWh) to be satisfied through the waste heat from power plants. This amount is less than the residual heat emitted by power plants located in the EU-28, at 19,608 PJ/year, which is mostly sent to cooling towers and not used for heating buildings as it is at too low a temperature (typically 30 °C for large thermal power plants) to be used economically. Currently, heating is mainly obtained from fossil fuels, which creates economic, environmental, and energy dependency problems. It should be noted that the total cost of imported energy in the EU-28 for the year 2013 was 405.8 billion euros, whereas the exported energy was barely 10 billion euros.

Chapter 1
Introduction

Energy and persistence conquer all things.
—Benjamin Franklin

The effect of the large cogeneration plants is identical to that of an electric heat pump in the sense that both use the electricity generated in a power plant to increase the temperature of the reject heat to a temperature at which this heat is useful in district heating networks. As a consequence, cogeneration can be considered as a "virtual heat pump," with the combined use of district heating networks and cogeneration being in practice considerably more effective than a domestic heat pump. The coefficient of efficiency (ratio of heat supplied per unit of electricity used) of existing cogeneration facilities is usually between 6 and 10 (higher coefficients of performance can be achieved if low-temperature district heating networks are used). In the case of an electric heat pump, its coefficient of performance varies, being higher in summer and lower in winter (the average throughout the year would be around 3).

As a consequence, the expenditures for primary energy and CO_2 emitted when the residual heat temperature of conventional thermal plants is increased through cogeneration are much lower than those incurred with other heat sources. District heating networks can cope with much of the industrial and domestic thermal loads in an economical way; it conflicts with reducing building loads through improved insulation which, although advisable, is unlikely to achieve the same level of CO_2 emission reduction as the combined use of district heating and co-generation networks. This is because domestic thermal loads are difficult to reduce, requiring, in any case, a high cost to achieve an adequate level of insulation and a reduction of these loads.

Obviously, heating, gas, or electric networks have certain factors of simultaneity, such that the load seen by the network is less than the sum of the peak loads of all buildings. On the contrary, individual options such as boilers, micro-cogeneration, or heat pumps have to be designed for the maximum theoretical loads of buildings and are therefore more costly. Therefore, there are considerable benefits from the

A. Colmenar-Santos et al., *District Heating and Cooling Networks in the European Union*, DOI 10.1007/978-3-319-57952-8_1

integration of heat, gas, and electricity networks particularly because of the ability of heat networks to store (much more economically than electrical storage) peak heat demands and absorb and smooth the fluctuating energy coming from renewable energies.

Taking into account their potential, the use of the above residual heat has immediate practical applications as it could foreseeably result in a considerable reduction in the polluting emissions resulting from the combustion of fossil fuels, a reduction of the primary energy used by the thermal power plants, and savings in net electricity consumption. This implementation, although technically feasible, poses institutional and financial challenges that cannot be solved without adequate measures.

Although the innovation proposed in this book is contrary to the competitive logic existing in most EU-28 Member States (as focal organisations will try to maintain their privileged status quo), the widespread use of district heating networks and cogeneration is not in itself a disruptive strategy since much of the existing infrastructure will still be usable. It is noteworthy that the combined use of district heating networks and cogeneration is potentially applicable to a large proportion of loads located within a "reasonable" distance of a thermal plant (typically between 60 and 70 km) and is economically advantageous, under some circumstances, for the southern EU-28 countries. Therefore, the proposed solution has a pan-European dimension in terms of energy challenges and markets.

Another factor to consider in relation to the implementation of the energy strategy proposed in this book is that the roadmap for the generalisation of the joint use of district heating networks and cogeneration networks typically has to cover decades to achieve widespread adoption (which means that there is a potential for improving energy efficiency over the years), since, by definition, infrastructural investments will require a high initial investment that must be adapted to the circumstances of each Member State. Although the technology proposed in this book is at a research stage in which it has been validated and tested in an operational environment (mainly in the Nordic countries), the existing institutional and financial barriers for most Member States are such that, often, the participation of potential relevant actors in the sector is complicated. It should be noted that district heating networks are a reliable technology, where the end-user only needs a consumption unit composed of heat exchangers, pumps, and valves to achieve a low annual maintenance cost (as it lacks a combustion chamber, chimney cannons, and fuel storage).

Also, such networks have sufficient intrinsic storage to cope with short cuts in cogeneration plants and peak load systems that can cover more severe cuts. On the other hand, and although its use is expected to decrease gradually, fossil fuels will, in all likelihood, continue to be a relevant energy resource for several decades more due mainly to the difficulty of replacing and building new power plants—so the proposed scheme in this book is a robust solution that is unlikely to be replaced.

There are some examples of how countries have managed to minimise their dependence on imported fossil fuels through the joint use of district heating and cogeneration networks. Denmark is an example of a country where a high level of implementation of this scheme (since the 1973 oil crisis) has enabled a progressive decline in fossil fuels imports and one of the lowest per capita primary energy usage. It should be noted that heating buildings through renewable electricity using the electricity grid (either through individual heat pumps or direct heating by resistance) presents practical difficulties due to the necessary investment to repower distribution networks and the need to increase the nominal rating of the power plants to cope with peak loads. On the other hand, domestic heat pumps, when connected to the low voltage grid, will incur in very high marginal losses, particularly in winter (when annual peak heat consumption occurs) as they are necessarily less efficient due to the lower coefficient of performance during this time of year.

With regard to large-scale electricity storage, it should be borne in mind that storing electricity from renewable sources for heating and cooling buildings involves an extremely costly investment. The storage system will have to cope with the intrinsic variability of many of the renewable resources plus the fact that there will be the problem of distributing electricity to the end-user through the electric distribution network, which, as already mentioned above, must in many cases be re-powered. This problem could be solved by using a much cheaper and available storage system, as would be the case with the combination of thermal storage and district heating networks.

To write this book, discussions were held with international experts (see section "Acknowledgements") in cogeneration thermal plants and district heating networks, who agreed that a much lower cost of heat and lower CO_2 emissions would be achieved if the district heating networks operated at low temperature (close to 75 °C for the flow and 30 °C for the return); however, there is an agreement among the experts that indicated that there is currently no demand for such schemes, which illustrates the lack of awareness of their possibilities (see Chap. 5). If possible, district heating networks and cogeneration should not use traditional temperatures (typically 120 °C), but the lower operating temperatures above, because the economic effectiveness of the scheme decreases. These lower temperatures are much more compatible with other renewable heat sources such as solar or geothermal. It is essential that the market is aware of the existence of these options and explore them in a greater degree of detail so as to stimulate demand.

The aim of this book is, therefore, to propose a practical formula through which to take advantage of part of the losses of the energy balance of the EU-28, which, according to the Sankey diagram of its energy balance (Fig. A-1), are in the aggregate of the 40,000 PJ and have a market value of about 630 billion euros,[1] assuming the price of a barrel of crude is at 100 dollars.[2] This low and usable calorific heat is to be used in the first instance for heating and cooling cities in

[1]For the oil barrel, an energy content of 35 MJ/litre, a volume of 159 litres, and the Euro/US dollar equivalence of 1.13 have been assumed.

[2]In spite of the collapse of the price of oil in February 2016, reaching a minimum of US $25, this has been considered a "typical" price of a barrel of crude oil (note that in June 2008 it reached US $145).

district heating (and cooling) networks and cogeneration—through which it would increase in temperature until it becomes usable.

As shown in Figure A-1, about 19 000 PJ of these losses originate as a by-product of power plants. This low heat is sent to cooling towers and generally not used for heating buildings because it is at too low a temperature to be economically used. This amount is considerably higher than the demand for heating for domestic and commercial loads in the EU-28, which is currently mainly supplied from fossil fuels.

Due to its cost competitiveness, its flexibility in terms of its ability to use renewable energy resources (such as geothermal or solar thermal) and fossil fuels (more specifically the residual heat from combustion), and the fact that the abovementioned losses in the EU energy balance can be easily integrated into district heating networks (which would not be the case in a "fully electric" future), if appropriate measures were proposed, district heating (and cooling) networks and cogeneration could become a key element for a future with greater energy security, while being more sustainable. It is precisely the proposal of an energy strategy for the EU-28, made possible by these appropriate measures and supported by detailed case studies, that the main objective (at the same time of achievement) resulting from the research is carried out in this book.

This book has been divided into seven chapters and annexes and has been drafted in such a way that, although each chapter can be read to some extent independently, has the necessary illation to facilitate its readability.

After this first introductory chapter, and to put into context the possible costs and energy savings calculated for the three cities under study in Chaps. 5, 2 assumes that around 50% of domestic and commercial heating demands of the EU-28 are met through the joint use of district heating networks and cogeneration. This scenario compares country by country with available heat, assuming that all conventional thermal plants are converted to cogeneration plants (an infeasible scenario).

Chapter 3 discusses institutional and financial barriers to the implementation of district heating and cogeneration networks in the EU-28. To this end, it points out that the existing barriers are not technological, and that the scheme of district heating networks/cogeneration is generally advantageous if it is evaluated at the national level, which is the reason for their failure largely due to institutional barriers.

Chapter 4 presents a detailed discussion of why there is no conflict in assessing the feasibility of schemes involving the combined use of cogeneration and district heating networks at a discount rate of 3.5% per year (when the private sector demands a much higher rate of return).

Chapter 5 provides an assessment of the total cost of the proposed scheme for three cities with representative EU-28 climatic conditions (Cartagena, Bristol, and Oldenburg) and compares it with heating (and in the case of Cartagena, also refrigeration) through gas and electricity.

Finally, as a documentary appendix, a section is reserved for the annexes where additional information will be shown to facilitate and illustrate the understanding of the main body of the book.

Chapter 2
District Heating and Cogeneration in the EU-28: Current Situation, Potential and Proposed Energy Strategy for Its Generalisation

The most common commodity is unrealised potential.
—Calvin Coolidge

Yearly, EU-28 conventional thermal generating plants reject a greater amount of energy than what ultimately is utilised by residential and commercial loads for heating and hot water. If this waste heat were to be used through district heating networks, given a previous energy valorisation, there would be a noticeable decrease in imported fossil fuels for heating. As a consequence, benefits in the form of an energy efficiency increase, an energy security improvement, and a minimisation of emitted greenhouse gases would occur. Given that it is not expected for heat demand to decrease significantly in the medium term, district heating networks show the greatest potential for the development of cogeneration. However, to make this happen, some barriers that are far from being technological but are mostly institutional and financial need to be removed.

The purpose of this chapter is to provide information on the potential of using waste heat from conventional thermal power plants (subsequently converted into cogeneration plants) in district heating networks located in the EU-28. For this, a preliminary assessment is conducted in order to show an estimate of the cost of adopting an energy strategy in which district heating networks are a major player of the energy mix.

From this assessment, it is possible to see that even though the energy strategy proposed in this book, which is based on a dramatic increase in the joint use of district heating networks and cogeneration, is capital-intensive and would require an annual investment of roughly 300 billion euros, its adoption would result in a reduction of yearly fuel expenses in the order of 100 billion euros and a shortening of about 15% of the total final energy consumption, which makes it of paramount interest as an enabler of the legal basis of the "Secure, Clean and Efficient Energy" future enacted by the EU-28 Horizon 2020.

A. Colmenar-Santos et al., *District Heating and Cooling Networks in the European Union*, DOI 10.1007/978-3-319-57952-8_2

2.1 Introduction

From the simplified energy balances performed annually by EUROSTAT (2013 is the last year for which data is available), conventional thermal power plants located in the EU-28 waste more energy in the form of residual heat than what finally residential and commercial buildings use for heating [1–3]. As is clear from Fig. 2.1, if this residual heat could be used previous energy valorisation from cogeneration in district heating networks, a noticeable decrease in the amount of imported fossil fuels and greenhouse gases emitted into the atmosphere would occur (at Appendix G, additional information on the relation between electricity and heat in a cogeneration plant is included), which would result in economic, environmental, and energy security benefits.

From Fig. 2.1, it is impossible to determine how much heat can be used. Many power plants have to be necessarily located far from the thermal loads of cities (e.g., coal power plants are usually located close to a port to stock). However, in a future scenario, it could be more beneficial, from an economic perspective, to place those plants in the vicinity of the cities. As they are environmentally friendly they can be located closer to them. The old plants should be dismantled when their useful life ends. This supposes an additional expense of transporting coal or lignite in order to be able to use waste heat from power plants [4].

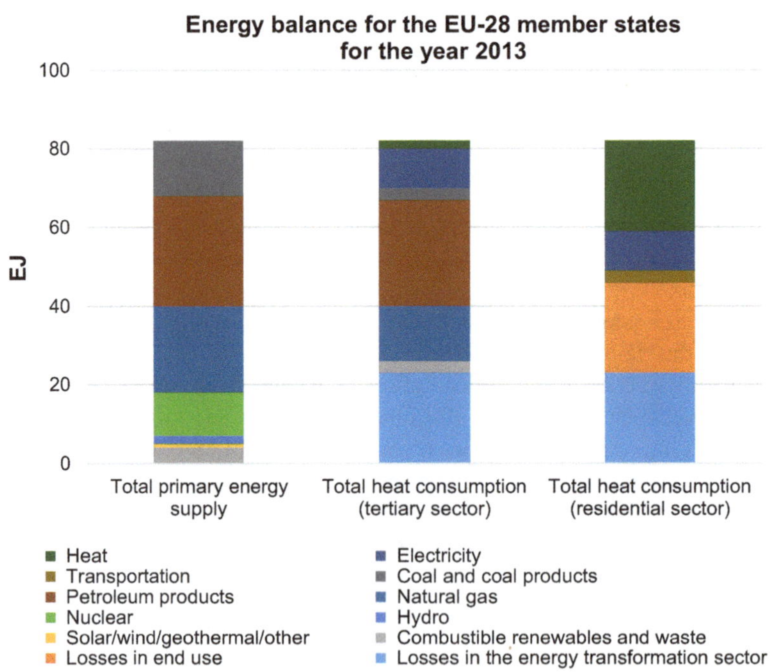

Fig. 2.1 Energy balance for the EU member states for the year 2013 [3]

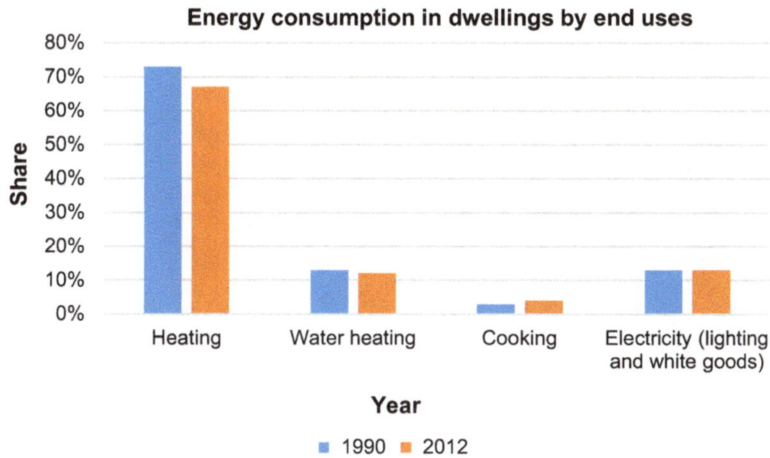

Fig. 2.2 Power consumption for households located in the European Union depending on the energy end-use *Source* Adapted from [38]

In Fig. A-1, Appendix A, it can be seen as another way of representing the energy balance (which again shows the importance of waste heat from power plants) [5].

As seen in Fig. 2.2, about 80% of the energy used in homes has as ultimate goal to provide heating and hot water; this means that, according to Fig. A-1, there would be potentially about 15,000 PJ/year (4100 TWh) to be satisfied through the waste heat from power plants (note that this amount is less than the residual heat emitted by power plants located in the EU-28, 19,608 PJ/year (Fig. A-1).

Finally, Fig. 2.3 shows the energy share for each district heating resource—it is possible to see that, for the EU, most of the heat comes from waste heat from power plants.

Despite these data, it should be noted that, due to the lack of energy matching throughout the year, as well as to the problem of the reject heat temperature (co-generation can conduct an energy valorisation of this low-temperature heat in an economical way), not all of these thermal loads can be satisfied by the waste heat from thermal power plants. This makes necessary to estimate their real potential including assumptions as close as possible to the reality and to conduct an economic and environmental assessment (reduction of greenhouse gas emissions) that includes a generalised spread of district heating networks using as an energy source waste heat from conventional thermal power plants in the EU-28, which will be conducted in this chapter.

The objective of this chapter is to propose an energy strategy which enables, through the generalisation of the joint use of district heating networks and cogeneration in the EU-28, the fulfillment of the legal basis of the H2020 "Secure, Clean, and Efficient Energy" goals in a feasible, reliable, economically advantageous, and sustainable way. The method proposed here uses a systematic and easy

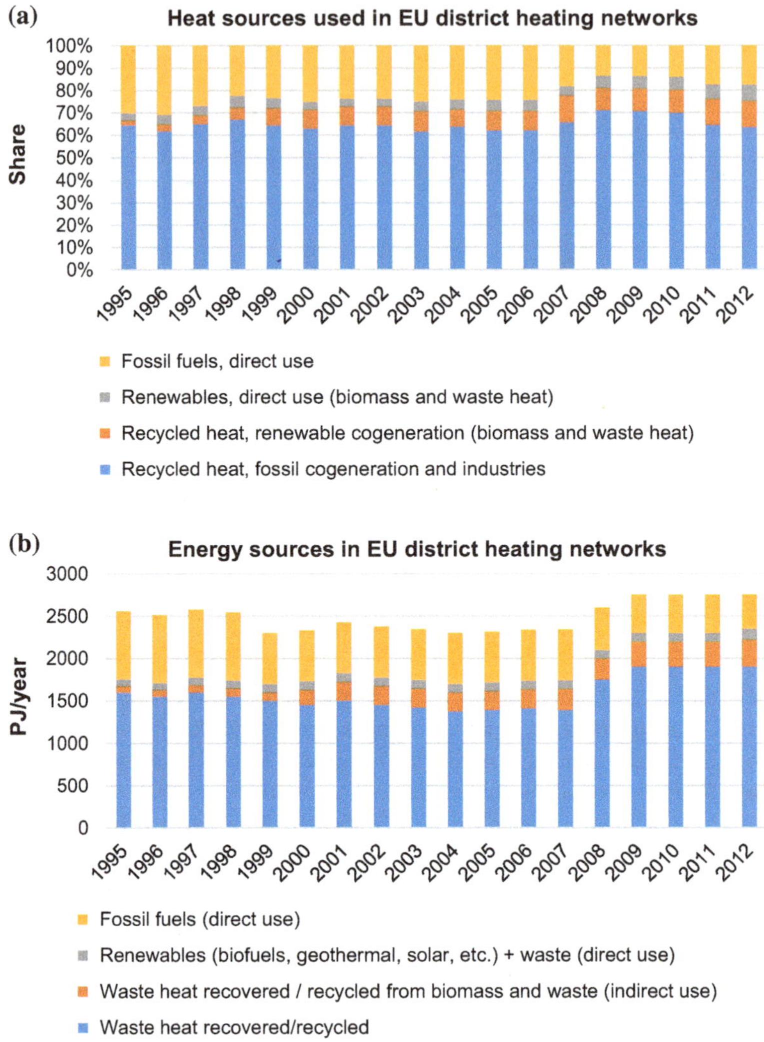

Fig. 2.3 Energy share (**a**) and generation for each district heating source (**b**) in the EU-28 [19, 39]

approach (neither strong mathematical skills nor non-widely used software is required) and manages to propose an energy strategy for such a big region as EU-28 based mostly on available public data (with the exception of the plant data obtained through the GESTIS database by using the ArcGIS software).

This is of great importance as other methods are network-oriented (district heating network); lack from massive surface applicability, being inadequate for such a big area as Europe; are limited to industrial sites; or present assessments which need a district heating network taxonomy, information not available in the vast majority of the cases. As a consequence, they are not useful for the purposes of this book. To our knowledge, there is not a single scientific paper which specifically addresses the potential of the joint use of thermal power plant reject heat into district heating networks across the European Union. Thus, we offer new insights into this matter.

The first section of this second chapter has made a first approximation of the energy balance of EU-28 Member States has been made and mentioned the potential use of reject heat from conventional thermal power plants. Section two will explain the methodology followed to develop the research carried out here; section three will show different studies evaluating the potential joint use of district heating networks and cogeneration for different regions; and section four will present different scenarios on the evolution of district heating networks and estimate the (economically feasible) potential of converting conventional thermal power plants into cogeneration plants.

The fifth section will be reserved for the discussion, and the sixth for the conclusions, which will show the implications (restrictions and benefits) of the generalisation of the joint use of cogeneration and district heating networks in the EU-28.

Finally, as additional information to this second chapter, annexes with in-formation on prices of district heating networks (Annex B), heat transportation (Annex C), heat distribution (Annex D), and thermal storage (Annex E) have been included. Furthermore, the case of Copenhagen as an example of a city that has been able to adequately combine the use of electricity, district heating, natural gas, and waste management (Annex H) has been presented.

2.2 Materials and Methods

To give documentary strength to this chapter, it was necessary in the first instance to consult more than 100 scientific publications and/or gray literature in the field of cogeneration and district heating networks. While not absolute, it is a representative sample of the state of the art of the subject under study.

Subsequently, and to determine the degree of matching between the above publications and the perception of experts from various members of the EU-28, a total of nine experts from seven countries of the European Union (see section "Acknowledgements") were consulted through questionnaires, email correspondence and/or telephone conversations.

Once the scientific/gray literature and the experts were consulted, the third step taken was to use various databases to estimate the potential for the joint utilisation of district heating networks and cogeneration. Among these databases, the most notable were EUROSTAT, GESTIS database based on Platts, and The European Pollutant Release and Transfer Register (European Environment Agency). The latter was essential for selecting those conventional ones potentially convertible in an economical way.

The fourth step taken was to introduce some restrictions on the type of plant and the distance from it to the thermal load (nearest town). To estimate the cost related to the heat demand, member states of the EU-28 were firstly divided into zones, with the countries of central and southern Europe assigned a cost on energy demand different from the one used only for the countries of southern Europe. This discrimination is based on the fact that, obviously, the costs of providing heat to the thermal loads decrease as demand increases.

Finally, and in order to know the cost of providing heat through district heating networks for different typical climatic zones of the EU-28, [6] was consulted, which constitutes a supporting feature to the review presented here. To the heat demand for each consumption located in Northern and Central Europe, a value of 2000 GWh/year was assigned, having been established the distance from the heat load at 75 km. Meanwhile, for those loads located in the Southern European countries, a type value of 750 GWh/year of heat and 400 GWh/year for cooling was established, being the distance in this case also 75 km. Note that these conditions are much more restrictive than those described in [6]; we have proceeded in this way because such cases, although based on real plants, have better characteristics for the conversion to cogeneration plants than those that would be considered as "typical".

Due to the fact that DHNs and DCNs lack a symmetrical end-use purpose (DHNs are usually aimed at areas where there are residential buildings within a city, while DCNs tend to be focused only on the central areas of the cities) and its proximity-to-load requirement, it is not possible to distribute cold appropriately from an economic perspective. District cooling is based on a difference of temperatures much smaller than district heating; see respectively in Appendix D and Appendix F additional information on the distribution of heat and cold in modern heating/cooling systems) [7]. It would be impractical to transmit the cold at a distance of 75 km, so for the countries of Southern Europe, it has been supposed the use of absorption heat units that use the heat from district heating networks for cooling, not being, therefore, necessary to use pipes for transporting cold at a large scale.

Some restrictions have been imposed, e.g. only those thermal power plants with a rated power of over 300 MW located within 75 km of urban centers of more than 100,000 inhabitants are considered as valid for conversion, and some data such as the following was found out:

- total installed capacity of thermal power plants using fossil fuels,
- total installed capacity of cogeneration,
- CHP generation percentage,
- CHP maximum installable capacity from thermal plants.

From this data, a feasible CHP capacity that would comply with the aforementioned restrictions is obtained and presented in the Results section. From these feasible convertible thermal power plants and from data obtained from the literature, usable heat, the investment needed, and fuel savings are obtained and shown in the Results section.

A summary of every stage of study and its relationship with the Results section would be as follows:

(1) Scientific publicatons and/or gray literature review.
(2) Experts in the field of cogeneration and district heating networks consultation.
(3) Databases (EUROSTAT, GESTIS, and The European Pollutant Release and Transfer Register) consultation.

   - The obtainment of conventional thermal combustion capacity, CHP electrical capacity, share of CHP electricity production, electricity generation from generation plants, heat generation from CHP plants, final total energy consumption, total district heating network sales, and percentage of citizens served by district heating, is immediate from EUROSTAT databases.
   - The methodology to assemble the data obtained from GESTIS and EPRTR was as follows:

      a. Plant data: GESTIS database based on Platts

         i. This data was accessed through the GESTIS database using the ArcGIS software.

      b. Plant coordinates: EPER/E-PRTR database

         i. It can be accessed through the European Pollutant Release and Transfer Register website. Then following steps were undertaken "Search the register" > "Facility level" > "Country" > "EU27" > "Region" > "All regions" > "Year" > "2013" > "Industrial activity" > "Energy sectors" > "Thermal power stations and other combustion industries" > "No sub-activities"

      c. City coordinates: Google Earth

         i. No description needed

(4) Calculation of maximum conversion of CHP by using total conventional thermal combustion capacity and CHP electrical capacity; heat to power ratio by using heat generation from CHP plants and electricity generation from CHP plants; and maximum annual additional heat output by using the maximum conversion of CHP calculated, the heat to power ratio, and supposing a load factor of 0.6.
(5) Restrictions on the type of plant (only conventional thermal plants with a rated power of over 300 MW are to be considered), distance (thermal plants shall be located at a distance below to 75 km of urban centers), population of urban centers (higher to 100,000 inhabitants) and heat demand in Northern and Central Europe (2000 GWh/year) and heat and cooling demand in Southern Europe (respectively 750 and 400 GWh/year) are applied.

(6) Calculation of DHNs feasible potential by using the maximum annual additional heat and the fact that, after the restrictions imposed in 5), about 50% of the installed capacity of all conventional thermal power plants comply with those restrictions; calculation of conversion and district heating costs by classifying countries into Regions (Norhtern and Central Europe and Southern Europe) and taking into account heating (and cooling) costs in those regions.
(7) Once DHNs' feasible potential is achieved and yearly savings in fuel are known, the cost of natural gas for domestic consumption is determined, and the usual fuel consumption per unit of heat supplied for boilers and cogeneration plants (for both cases separately) are assumed. Through the cost difference between the two technologies, fuel savings are obtained.
(8) Finally, the results from previous steps (1–7) are analysed, and an energy strategy for the generalisation of district heating networks and cogeneration is proposed, achieving, consequently, the objectives of the research.

2.3 Theoretical Background

To know the true potential of any technology, in this case, the joint use of cogeneration and district heating networks, it is necessary to perform an assessment of its economic feasibility. In the existing literature, it is possible to find a relatively high volume of gray literature and scientific publications that carry out feasibility studies relating to the implementation of district heating networks.

One of the first was conducted in 1979 by the Ministry of Energy of the United Kingdom [8], where it was indicated that the joint use of cogeneration and district heating networks was competitive already at that time from an economic perspective in comparison to the fuels used to satisfy the heating demand (the largest infrastructural costs were offset by lower operating costs). This study recommended the development of a program that would implement district heating networks associated with cogeneration plants. Although this publication saw the light more than three decades ago, the results remain equally valid because, firstly, the technical installation of the district heating networks have improved and the costs have decreased [9] and, secondly, comparatively fuel-related costs have increased more than the cost of the necessary infrastructure [10].

A computerised methodology to estimate the potential of the joint use of cogeneration and district heating for various discount rates was used [11]. As shown in Table 2.1, it is only in the UK with a peak demand of 60 GW [12] that there is a potential of 33 GW for this technology for a discount rate of 3.5%. Due to similar weather conditions, this study is representative for much of northern and central Europe. As is clear from Table 2.1 and Fig. 2.4, the choice of the discount rate is crucial to the feasibility of the project.

In [12, 13], the possibility of using cogeneration and district heating networks is studied again, reaching similar conclusions. As is clear from these publications, the

Table 2.1 Potential joint use of cogeneration and district heating networks in the UK for various discount rates [11]

CH/CHP potential	Units	Discount rate (%)		
		3.5	6	9
Total net CH/CHP potential for UK	MWe	33,125	21,517	75
Total electricity produced	–	6897	4204	46
Total heat sold	GWh	189,472	123,119	518
Primary energy saving	GWh	230,358	149,686	630

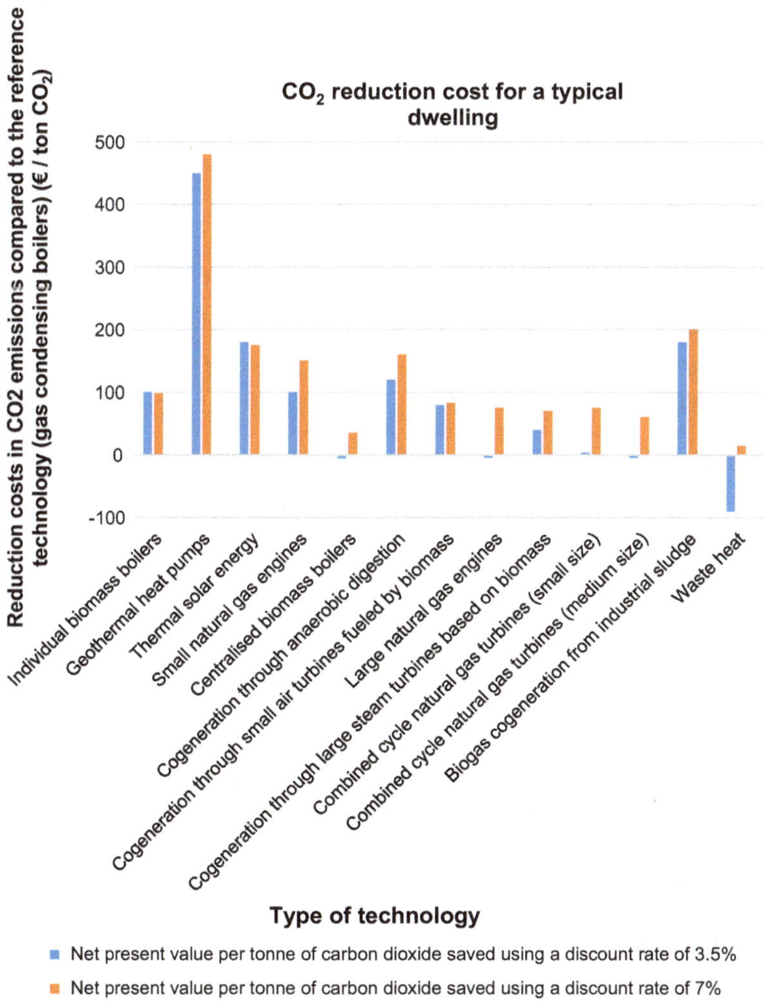

Fig. 2.4 The cost of CO_2 reduction for the typical UK dwelling [15]

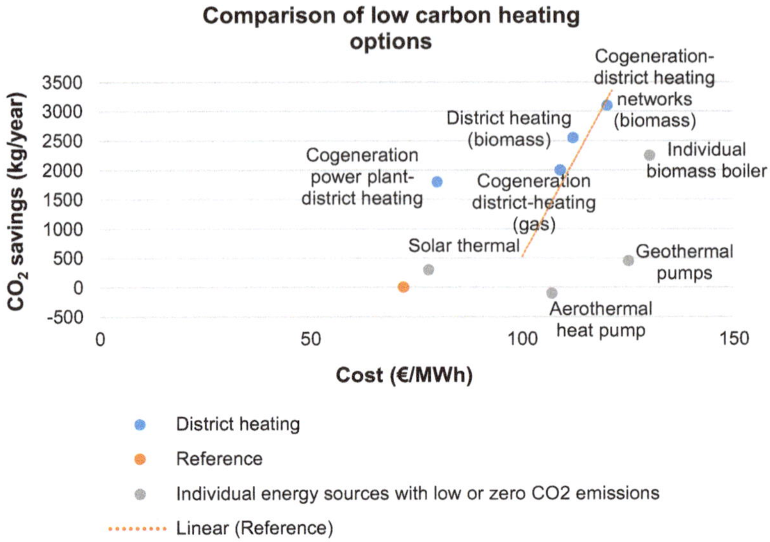

Fig. 2.5 Comparison of different energy solutions with low emissions of carbon dioxide [16]

joint use of these technologies provokes one of the greatest savings, in terms of CO_2 emissions, at one of the lowest costs per not emitted tonne [13, 14] (Figs. 2.4, 2.5).

At a European level, [15] confirms the possibility of stopping to emit 400 $MtCO_2$ per year, which would mean a reduction of 9% compared to current emissions, if the penetration of district heating and cooling were to be doubled.

2.4 Results

Given its economic and environmental benefits, the joint use of cogeneration and district heating networks should experience significant growth in coming years (Fig. 2.6) [16]. For district heating networks to achieve a penetration of over 90% in a big city, several decades may elapse, in the case of Copenhagen, it took more than 40 years [17], depending largely on the commitment of the authorities [18]. Appendix H includes additional information concerning the case of Copenhagen. Note that the actions to take for the widespread introduction of district heating networks from the market entry stage (going through the market penetration phase) to the stage of market maturity will depend, in any case, on the casuistry of each location; therefore, an adaptation of the measures to be taken depending on specific circumstances will be necessary.

To know the potential costs and energy savings, it has been supposed as a target that 50% of all domestic and commercial heating loads from the EU-28 can be satisfied from waste heat from conventional power stations and district heating

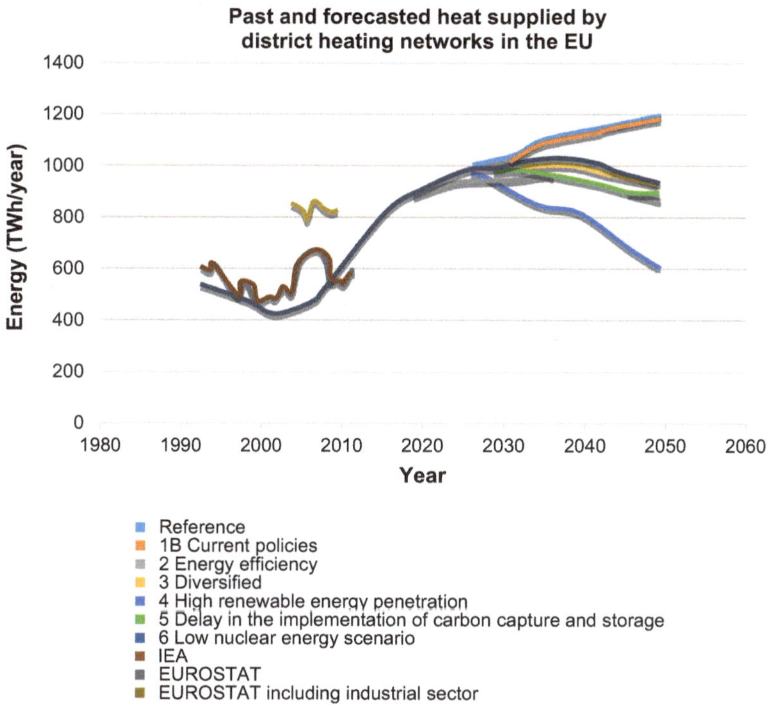

Fig. 2.6 Different scenarios for the heat supplied by district heating networks and cogeneration plants located in the EU until 2050 [16]

networks. These results are compared country by country assuming power plants are converted into cogeneration plants.

The estimate has been made as follows:

1. Table 2.2 gives the total installed capacity of thermal power plants using fossil fuels by member state in column A.
2. Column B shows the total installed capacity of cogeneration per member state (Table 2.2).
3. Column C presents the combined heat and power (electrical) generation percentage from each member state (Table 2.2).
4. Column D (Table 2.2) shows the installable capacity in the EU-28, assuming that 100% of conventional thermal power plants that use fossil fuels are converted into cogeneration plants (although not all the heat is used because the heat generated is higher than the actual demand).

Column E (Table 2.3) represents the additional energy obtained from converting all conventional power plants placed in the EU-28 into cogeneration plants, as shown in column D (Table 2.2). It is supposed that once the conventional thermal power plant has been converted, LF is 0.6. HPR calculations were carried out, taking into account columns A and B (Table 2.3), which correspond to the statistics

Table 2.2 Installed capacity per member state to achieve full conversion of conventional thermal power plants in cogeneration plants

	A	B	C	D
	Total conventional thermal combustion capacity [41] (GW)	CHP electrical capacity [42] (GW)	Share of CHP in electricity production [42]	Maximum conversion of CHP (GW)
	2013	2013	2013	Vision
AT—Austria	6.3	4.4	14.4%	1.9
BE—Belgium	7.6	2.3	15.2%	5.3
BG—Bulgaria	4.6	1.2	8.5%	3.4
CY—Cyprus	1.6	0.0	1.4%	1.6
CZ—Czech Republic	10.5	4.6	13.7%	5.9
DE—Germany	82.0	27.0	12.4%	55.0
DK—Denmark	7.6	5.7	50.6%	1.9
EE—Estonia	2.6	0.5	9.3%	2.1
ES—Spain	49.8	4.0	8.5%	45.8
FI—Finland	8.4	6.3	34.5%	2.1
FR—France	23.0	5.5	2.4%	17.5
GR—Greece	10.7	0.6	3.4%	10.1
HU—Hungary	5.8	1.6	13.4%	4.2
HR—Croatia	1.7	0.7	12.6%	1.0
IE—Ireland	6.0	0.3	7.8%	5.7
IT—Italy	70.6	7.6	12.7%	63.0
LT—Lithuania	2.8	1.2	35.0%	1.6
LU—Luxembourg	0.5	0.5	14.7%	0.0
LV—Latvia	1.2	1.0	38.3%	0.2
MT—Malta	0.6	0.0	0.0%	0.6
NL—Netherlands	21.4	9.2	34.5%	12.2
PL—Poland	28.2	8.3	15.9%	19.9
PT—Portugal	6.4	1.4	13.8%	5.0
RO—Romania	9.0	2.2	11.2%	6.8
SE—Sweden	6.6	4.5	10.2%	2.1
SI—Slovenia	1.2	0.3	7.1%	0.9
SK—Slovakia	2.8	4.4	77.0%	0.0
UK—United Kingdom	59.0	6.2	5.5%	52.8
EU-28	438.5	111.5		328.6

Source Adapted from [41, 42]

Table 2.3 Calculation of the annual heat obtained from the conversion of conventional thermal power plants located in the EU-28

	A	B	C	D	E
	Electricity generation from CHP plants [43] (TWh)	Heat generation from CHP plants [43] (TWh)	Heat to power ratio	Maximum conversion of CHP (GW)	Maximum annual additional heat output (TWh)
	2013	2013	2013	Vision	Vision
AT—Austria	9.9	30.8	3.1	1.9	31.0
BE—Belgium	12.7	7.6	0.6	5.3	16.7
BG—Bulgaria	3.7	11.2	3.0	3.4	53.6
CY—Cyprus	0.1	0.0	/	1.6	8.4
CZ—Czech Republic	12.0	33.6	2.8	5.9	86.8
DE—Germany	78.7	181.7	2.3	55.0	664.9
DK—Denmark	17.6	28.7	1.6	1.9	16.0
EE—Estonia	1.2	3.5	2.9	2.1	32.0
ES—Spain	24.1	48.6	2.0	45.8	481.4
FI—Finland	21.3	69.8	2.9	2.1	32.0
FR—France	14.0	41.8	3.0	17.5	275.9
GR—Greece	2.0	2.9	1.5	10.1	79.6
HU—Hungary	3.9	7.5	1.9	4.2	41.9
HR—Croatia	1.7	3.7	2.2	1.0	11.6
IE—Ireland	2.0	3.4	1.7	5.7	51.0
IT—Italy	36.7	59.1	1.6	63.0	529.8
LT—Lithuania	1.7	4.3	2.5	1.6	21.0
LU—Luxembourg	0.4	0.9	2.3	0.0	0.0
LV—Latvia	2.4	3.1	1.3	0.2	1.4
MT—Malta	0.0	0.0	/	0.6	3.2
NL—Netherlands	34.8	60.5	1.4	12.2	89.8
PL—Poland	26.1	71.5	2.7	19.9	282.4
PT—Portugal	7.2	19.0	2.6	5.0	68.3
RO—Romania	6.6	16.1	2.4	6.8	85.8
SE—Sweden	15.6	45.9	2.9	2.1	32.0
SI—Slovenia	1.2	3.0	2.5	0.9	11.8
SK—Slovakia	22.2	7.7	0.3	0.0	0.0
UK—United Kingdom	19.7	39.6	2.0	52.8	555.0
EU-28	390.1	789.4	/	328.6	3563.3

Source Adapted from [42]

available from EUROSTAT for the main power producers and not for the one of the auto-producers, and assuming that the heat of a CHP plant will feed district heating networks.

We have proceeded in this way due to the fact that:

1. the vast majority of non-cogeneration thermal power plants belong to large power producers and not to auto-producers
2. and that the uncertainty of connecting the waste heat from a power plant to a district heating network is much lower than the one that would be obtained if industrial waste heat were to be used, where other factors, such as the possibility that a given industry remains at least 60 years working in a given location, should be evaluated.

Because district heating networks always will have availability restrictions, this book has centered on HHW requirements for tertiary and residential sectors. An analysis of the technical potential of the district heating networks is shown in Table 2.4. As indicated earlier in this section, it has been assumed that district heating networks can supply roughly 50% of all heating requirements, domestic and commercial ones. In Table 2.4, it is possible to see that it has been supposed that the potential feasible for district heating networks (column E) is 50% of the maximum theoretical heat, a scenario in which all domestic and commercial loads have access and are satisfied through district heating networks (column C).

Finally, the costs related to the conversion of conventional thermal power plants into cogeneration plants and the investment required for the expansion of the district heating networks are estimated in column G (Table 2.4). Expansion costs include costs relating to both heat transport as those incurred in the district heating network itself. It should be noted that the cost of network expansion will vary depending on the region under analysis. Locations where AHDD is high incur smaller expenses (for district heating) than those for regions with a low heat demand, where a lower number of buildings will need to be connected to the district heating network. For those projects located in northern and central Europe, denoted respectively under the "N-C" acronym (see column F of Table 2.4), it has been assumed that they incur an annual heat demand cost of 45 €/GJ [19]. The costs of providing heat and cold jointly to those projects located in southern European countries, denoted by the acronym "S" (column F of Table 2.4), were assumed to be 60 €/GJ [6]. Note, however, that carrying out an appropriate assessment of the factual potentiality of associating current thermal generating plants to DHNs (district heating networks) requires a detailed analysis of each particular case, in which various techno-economic parameters are taken into account–see [6]. That is an extra feature to this book, as this analysis is beyond the scope of this book.

As can be seen in column C (Table 2.4), about 12% of the citizens of the EU-28 have access to district heating networks, with significant differences between countries. It is possible to find cases where the contribution of district heating networks to heat demand is virtually zero, as in the case of Spain, and others where the weight of district heating networks is high (nearly 2 out of 3 Danish have access

Table 2.4 Analysis of the technical potential of district heating networks

	A	B	C	D	E	F	G
	Final total energy consumption [3] (PJ)	Total district heating networks sales [44] (PJ)	Percentage of citizens served by district heating [44, 45]	Maximum annual additional heat (PJ)	DHNs feasible potential (PJ)	Region	Conversion cost + DHNs cost [6] (annual cost of investment in milliards of €)
	2013	2013	2013	Vision	Vision		Vision
AT—Austria	1170.2	80.7	24%	111.6	55.8	N–C	2.4
BE—Belgium	1457.1	21.0	3%	60.1	30.1	N–C	1.4
BG—Bulgaria	367.2	18.0	18%	193.0	96.5	N–C	4.2
CY—Cyprus	67.6	0.0	0%	30.2	15.1	S	1.0
CZ—Czech Republic	1000.0	89.4	38%	312.5	156.3	N–C	6.8
DE—Germany	9096.0	254.8	12%	2393.6	1196.8	N–C	52.6
DK—Denmark	594.3	105.5	63%	57.6	28.8	N–C	1.4
EE—Estonia	120.2	23.0	62%	115.2	57.6	N–C	2.6
ES—Spain	3397.0	0.0	0%	1733.0	866.5	S	52
FI—Finland	1030.6	114.2	50%	115.2	57.6	N–C	2.6
FR—France	6366.3	96.1	7%	993.2	496.6	N–C	20.2
GR—Greece	642.2	1.0	0%	286.6	143.3	S	8.6
HU—Hungary	628.8	31.0	15%	150.8	75.4	N–C	3.4
HR—Croatia	243.4	9.7	10%	41.8	20.9	N–C	1.0
IE—Ireland	449.6	0.1	0%	183.6	91.8	N–C	4.0
IT—Italy	4969.6	33.1	6%	1907.3	953.7	S	57.2
LT—Lithuania	198.3	27.1	57%	75.6	37.8	N–C	1.6
LU—Luxembourg	172.9	2.0	6%	0.0	0.0	N–C	0.0

(continued)

Table 2.4 (continued)

	A	B	C	D	E	F	G
	Final total energy consumption [3] (PJ)	Total district heating networks sales [44] (PJ)	Percentage of citizens served by district heating [44, 45]	Maximum annual additional heat (PJ)	DHNs feasible potential (PJ)	Region	Conversion cost + DHNs cost [6] (annual cost of investment in milliards of €)
	2013	2013	2013	Vision	Vision		Vision
LV—Latvia	161.4	21.5	65%	5.0	2.5	N–C	0.2
MT—Malta	21.0	0.0	0%	11.5	5.8	S	0.4
NL—Netherlands	2141.8	26.1	4%	323.3	161.7	N–C	7.2
PL—Poland	2655.0	248.7	53%	1016.7	508.4	N–C	22.4
PT—Portugal	663.5	9.0	2%	245.9	123.0	S	7.4
RO—Romania	911.0	54.0	23%	308.9	155.0	N–C	6.8
SE—Sweden	1322.8	176.0	52%	115.2	57.6	N–C	2.6
SI—Slovenia	200.9	7.7	15%	42.5	21.3	N–C	1.0
SK—Slovakia	454.9	82.7	35%	0.0	0.0	N–C	0.0
UK—United Kingdom	5712.2	41.5	2%	1998.0	999.0	N–C	44.0
EU-28	46 214.5	1573.9	11.8%	12,827.9	6414.9		315.0

Source Adapted from [3, 6, 44, 45]

to this technology). A historical of the heat price provided by the district heating networks, the contribution of cogeneration to produce electricity, and the percentage of citizens who have access to district heating networks is presented in Figs. B-1, B-2, and B-3 in Appendix B section.

It is possible to see that the length of the pipes for transporting heat from the thermal power station located at Wilhelmshaven (757 MW) to the city of Oldenbourg is of more than 60 km and that this conversion would be feasible from an economic point of view. So it can be assumed conservatively that about 50% of all the electric power allocable to conventional thermal power plants is located within a reasonable distance to convert these plants in cogeneration plants [6]. This percentage has been obtained—from the database (E-PTR) [20]—after imposing as constraints that only will be valid for conversion to cogeneration plants those conventional thermal power plants with a rated output of over 300 MW placed at a distance of less than 75 km from towns of more than 100,000 inhabitants, which results in about 23% (this represents a 56% of the installed capacity of this technology) of all conventional thermal power plants not yet converted to cogeneration plants placed in the EU-28 (in 2013) to comply with these restrictions (see Table 2.5).

According to Table 2.4, if all potentially convertible thermal power plants, those with a rated power of over 300 MW located within 75 km of urban centers of more than 100,000 inhabitants, were converted into cogeneration plants and they would have access to district heating networks, about 6400 PJ of heat could be used, with a low carbon footprint, if an investment of some 315 billion euros will be made. After having consulted the database [20], it was found that in most countries, the limiting factor is, rather than lack of heat loads, the availability of thermal plants. However, in some locations, it was possible to observe the opposite, as groups of thermal power plants located relatively remote or at uneconomical distances to transport heat happened.

To get a clearer idea of the economic benefits of investing in district heating networks, let us assume that fuel consumption per unit of heat supplied is 1.1 in the case of boilers [21], being meanwhile 0.125 in the case of cogeneration plants [21]. Assuming that the cost of natural gas for domestic consumers was about 15 €/GJ for the second half of 2014 [22], yearly savings of 93.6 billion euros would be achieved only in fuel for an annual investment 315 billion euros.

Table 2.5 Conventional thermal power plants (not converted to cogeneration power plants) with a rated output of over 300 MW placed at a distance of less than 75 km from towns of more than 100,000 inhabitants

Number of conventional thermal plants (unconverted) near urban centers (distance <75 km and P > 300 MW)	Total conventional thermal power plants (unconverted) near urban centers (GWe)	Total number of conventional thermal power plants
393	246.5	1690

Source Database of E-PTR [20]

This estimate is only intended to give context about the treated magnitudes; in any case, it needs further evaluation case by case to debug this rough approximation; see [6]. It should be noted that, obviously, this amount is in itself an upper limit, which is subject to considerable uncertainty and that some of the conventional thermal plants that here have been supposed to be convertible eventually will not. However, if a greater commitment at the European level in promoting the joint use of district heating networks and cogeneration would be obtained, in the long term, it would be possible to approach the upper limit as new cogeneration power plants were built close to the thermal loads. Certainly, the barriers that the joint use of cogeneration and district heating faces are not technological but institutional and financial; see [23] to know these barriers in detail.

2.5 Discussion

As in most of the district heating networks of the member countries of the European Union that belonged to the former USSR and in the early stages as compared to the rest of the EU, it was common practice to use auxiliary boilers that only generated heat without there being at first a willingness to convert them into cogeneration plants [24–27]. Commonly, they used cheap or subsidised hydrocarbons [28] that had difficulties in burning [29]. With the passage of time, a number of auxiliary boilers from some cities were connected to other thermal generating plants that could use their reject heat. In Prague, what really triggered the connection of the Poříčí II power plant with the associated desulfurised emissions to the existing district heating network, even taking into account that the pipes carrying steam and hot water should have a length of about 40 km each one, was the realisation of the fact that there was a need to desulfurise emissions from plants that burned dirty fuels [30].

It might be thought that the joint use of cogeneration and district heating networks would only be profitable in the Northern European states due to their cooler climate and greater thermal requirements, rather than be consequently applicable to the rest of the EU-28. However, the fact that homes in these countries have better insulation has as a consequence that the specific thermal demands are lower (Fig. 2.7). Moreover, it is unlikely that the thermal loads in buildings will decrease so that the joint use of cogeneration and district heating networks becomes non-viable [31].

On the other hand, usual reductions of heat losses in buildings are not as high as expected in the first instance (the users to see an improvement in the efficiency of an energy service that leads to a reduction of the actual cost of the service tend to consume more, causing the known *rebound effect*) [32]. This effect, although difficult to quantify (but real), might result in a decrease in thermal load reductions of between 10 and 30% [32]. The vision of the not or, at least, modest changes in heat demand is based on forecasts conducted by each of the member states. As an example, it is presented the case of Ireland (Table 2.6), where it is possible to see

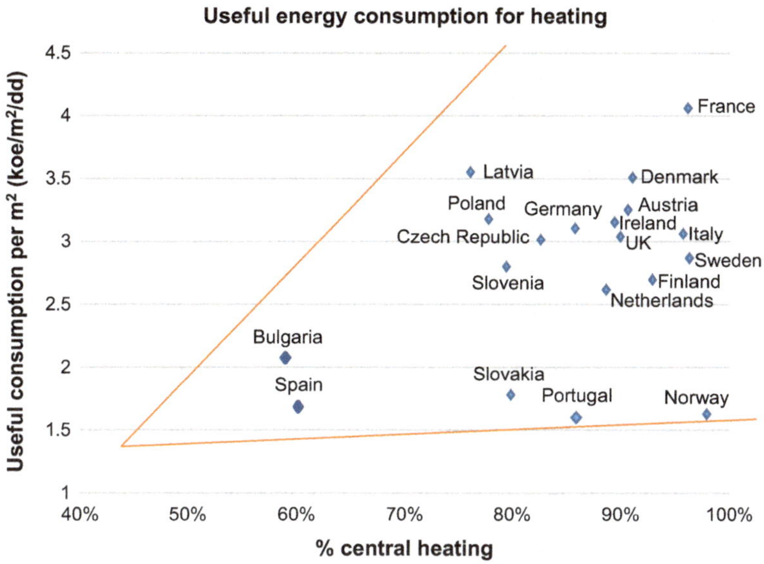

Fig. 2.7 Useful energy consumption for heating in equivalent kilograms of oil per m^2 and heating days with respect to the percentage of centralised heating for some EU member states [40]

Table 2.6 Thermal energy consumption by sector for Ireland (2008–2020) [33]

	Total final demand (ktoe)			Growth (%)	Average annual growth (%)			Sectoral shares (%)		
	2008	2012	2020	2008–2020	2008–2020	2008–2012	2012–2020	2008	2010	2020
End-use										
Industrial	139	148	150		0.6	1.5	0.2	2.5	2.8	2.6
Residential	44	43	41		−0.5	−0.5	−0.5	0.8	0.8	0.7
Services	15	14	14		−0.5	−1.0	−0.02	0.3	0.3	0.2
Total renewable energies	198	205	205	3.7	0.3	0.9	0.0			
Total consumption	5502	5209	5718	4.0	0.3	−1.3	1.2			
Heating from renewable energies								3.6	3.9	3.6

that for the period 2008–2020, it is expected to increase the thermal energy consumption by only a 0.3% [33], very much in line with other countries EU-28.

Furthermore, there are several reasons why it is easier to increase generation from renewable sources (particularly wind) than from cogeneration. In the EU-28, wind energy (along with hydroelectric power) is the dominant energy source from renewable origins [34, 35], being necessary in this case only that the promoter locates a location and would obtain the various needed permits. Because there are usually many locations where to deploy a project, there are high chances that the same is carried out. By contrast, cogeneration is undoubtedly much more restricted because it inevitably has to be used in cities and industries—obviously, the difficulties associated with implementing a district heating network to an entire city are of a different order than those found in obtaining permission for implementing a wind farm.

The cogeneration plants are also subject to more restrictions than those plants that produce heat and electricity separately. This is because although the electricity is cheap and easy to transport over long distances (in Appendix C section it is possible to find additional information concerning large-scale heat transportation), heat needs to be generated locally and can only be transported to much shorter distances. Consequently, a heat-generating unit will always be connected to the end user; meanwhile, electricity will feed to the electric grid in general. In other words, the cogeneration plant will have to compete in the heat market with, for example, heat boilers or heat pumps, and in the electricity market with more flexible power plants [36].

Despite these restrictions, it is to be emphasised that a higher CHP and DHNs generalisation (and their associated thermal storage; see Appendix E section) is of vital importance in reducing overall energy consumption in the EU-28, which obviously will impact CO_2 emissions. This view has been the one that has enabled Denmark, along with increased energy requirements of buildings, to prevent an increase in its energy consumption over the last 30 years despite having experienced an increase of 70% of GDP during this period [37].

Literature findings are certainly inconclusive in what district heating networks and cogeneration potential are respected at a pan-European level. In fact, even though some studies have been undertaken for specific countries or regions, scientific literature has not given the same attention to the potential and the economic and environmental benefits resulting from the implementation of an energy strategy based on a massive increase of the joint use of district heating networks and cogeneration in the EU-28. As a result, and due to the aforementioned lack of attention, a comparison between this book and other scientific publications is, at least, problematic (in particular, and for Southern Europe countries, attention given has been virtually zero). However, it is possible to indicate that one of the main strengths of this book is that, by using a simple methodology, its approach and perspective provide both novel results for the scientific literature and new insights to the European Union as a whole about the true potential of this underexploited technology. Paradoxically, this strength of the simple methodology is at the same time a shortcoming due to inaccuracies in the district heating and cogeneration potential assessment, and further research on this issue would be interesting for refinement purposes. Finally, and as another research extension, it would be interesting to evaluate the applicability of other reject heat from other types of thermal power plants such as biomass and nuclear ones for district heating purposes.

2.6 Conclusions

About half of the installed electric capacity in the EU-28, from conventional thermal power plants, is located at an appropriate distance to be able to convert these plants into cogeneration plants and to use their wasted heat in district heating networks. Taking into account the maximum penetration that is economically feasible, from an annualised investment in infrastructures of 315 billion euros, the costs, assignable to fuel, would be reduced by 95 billion euros per year and would save about 6400 PJ of primary energy. This represents about 15% of the total final energy consumption in the EU-28 in 2013 (46 214.5 PJ). All these savings are achieved from a by-product, such as the waste heat from conventional power plants, currently not utilised in order to meet any thermal load due to their low temperature (about 30 °C). If the energy vision, proposed in this book, were implemented, it would quadruple (from 12 to 50%) the access of the citizens of the EU-28 to the technology of district heating networks.

This estimate has important political implications as a generalisation of the joint use of cogeneration plants to raise the aforementioned temperature, and district heating networks would entail some economic, environmental, and energy security benefits that are necessary for the fulfillment of the legal basis of the H2020 Societal Challenges "Secure, Clean, and Efficient Energy" in a sustainable way. It has been consequently justified that, definitely, the barriers that the proposed new vision face are not of a technological nor of an economic nature but are institutional and financial.

It is worth to remark the forecast that, if a policy, committed to the development of district heating networks and cogeneration, is obtained, it would imply that over the next decades more and more power plants will be built in proximity to thermal loads and decommission those located remotely. Finally, and as a possible research extension, it would be interesting to consider reject heat from other thermal power plants, such as biomass and nuclear ones. Even though the latter ones should be considered cautiously due to their typical locations (normally far from heat loads), rejected heat from nuclear power plants might represent a significant increase in the joint use of CHP and DHNs, as currently they represent about 30% of the electricity provided in the EU-28.

References

1. Eurostat. Extra-EU28 imports and exports of energy products. Luxembourg: Eurostat. 2013. http://ec.europa.eu/eurostat/statistics-explained/index.php/File:Extra-EU28_imports_and_exports_of_energy_products,2013.png. Accessed 22 Oct 2015.
2. Ecoheatcool and Euroheat & Power. The European heat market: final report. Brussels: Euroheat & Power; 2006.
3. Eurostat. Simplified energy balances—annual data. Luxembourg: Eurostat. http://www.ec.europa.eu/eurostat/en/web/products-datasets/-/NRG_100A. Accessed 3 Oct 2015.

4. The Connecticut Academy of Science and Engineering. A study of the feasibility of utilizing waste heat from central electric power generating stations and potential applications. Hartford: The Connecticut Academy of Science and Engineering; 2009.
5. IEA. World balance. Paris: IEA. http://www.iea.org/sankey/. Accessed 22 Sept 2015.
6. Colmenar-Santos A, Rosales-Asensio E, Borge-Diez D, Collado-Fernández E. Evaluation of the cost of using power plant reject heat in low-temperature district heating and cooling networks. Appl Energy. 2016;162:892–907.
7. Boysen H. Hydraulic balance in a district cooling system. Nordborg: Danfoss District Energy; 2003.
8. Combined Heat and Power Group. Combined heat and electrical power generation in the United Kingdom: report to the secretary of state for energy. London: H.M. Stationery Off; 1979.
9. Euroheat & Power. District heating in buildings. Brussels: Euroheat & Power; 2011.
10. International Energy Agency. Cogeneration and district energy. Paris: International Energy Agency; 2009.
11. Department for Environment Food and Rural Affairs. Analysis of the UK potential for combined heat and power. London: Queen's Printer and Controller of HMSO; 2007.
12. Osborne M. Making electricity transmission smarter. Cambridge: Cambridge University Energy Network; 2013.
13. Kelly SJ, Pollit MG. An assessment of the present and future opportunities for combined heat and power with district heating (CHP-DH) in the United Kingdom. Energy Policy. 2012;38 (11):6936–45.
14. Woods P, Turton A. Smart heat grids—the potential for district heating to contribute to electricity demand management to facilitate renewable and nuclear electricity generation. Abingdon: The Solar Energy Society Conference C92; 2010.
15. Werner S. Benefits with more district heating and cooling in Europe. Rome: 20th World Energy Conference; 2007.
16. Aalborg Universitet. Heat roadmap Europe 2050. Aalborg: Aalborg Universitet; 2013.
17. DHC + Technology Platform. District heating & cooling: a vision towards 2020–2030–2050. Brussels: DHC + Technology Platform; 2009.
18. Thornton R. Copenhagen's district heating system: recycling waste heat reduces carbon emissions and delivers energy security. Westborough: International District Energy Association; 2009.
19. Euroheat & Power. District heating and cooling–statistics. Brussels: Euroheat & Power; 2015.
20. E-PTR. The European Pollutant Release and Transfer Register. Copenhagen: European Environment Agency European Environment Agency. http://prtr.ec.europa.eu/FacilityLevels.aspx. Accessed 3 Oct 2015.
21. Riddle A. District energy & smart energy grids experience from Denmark. London: Rambøll Energy; 2013.
22. Eurostat. Gas prices for domestic consumers—bi-annual data. Luxembourg: Eurostat. http://ec.europa.eu/eurostat/en/web/products-datasets/-/NRG_PC_202. Accessed 3 Oct 2015.
23. Colmenar-Santos A, Rosales-Asensio E, Borge-Diez D, Mur-Pérez F. Cogeneration and district heating networks: measures to remove institutional and financial barriers that restrict their joint use in the EU-28. Energy. 2015;85:403–14.
24. Martinot E. Investments to improve the energy efficiency of existing residential buildings in countries of the former Soviet Union. Washington: World Bank; 1997.
25. Danish Board of District Heating. District heating history. Frederiksberg: Danish Board of District Heating. http://dbdh.dk/district-heating-history/. Accessed 19 Feb 2014.
26. Gaigalis V, Markevicius A, Katinas V, Skema R, Tumosa A. Analysis of energy transition possibilities after the decommission of a nuclear power plant in Ignalina region in Lithuania. Renew Sustain Energy Rev. 2013;24:45–56.
27. Streimikiene D, Ciegis R, Grundey D. Promotion of energy efficiency in Lithuania. Renew Sustain Energy Rev. 2008;12:772–89.

28. Biomass Cogeneration Network. Current situation on CHP and biomass CHP in the national energy sector. Pikermi: Biomass Cogeneration Network; 2001.
29. Alternative Petroleum Technologies. Fuel oil emulsions. Reno: Alternative Petroleum Technologies; 2013. http://www.altpetrol.com/en/2d-pd-foe.html. Accessed 19 Feb 2014.
30. CEZ Group. The Poříčí power stations. Praga: CEZ Group; 2014. http://www.cez.cz/en/power-plants-and-environment/coal-fired-power-plants/cr/porici.html. Accessed 19 Feb 2014.
31. Wan KKW, Li DHW, Liu D, Lam JC. Future trends of building heating and cooling loads and energy consumption in different climates. Build Environ. 2011;46(1):223–34.
32. Sorrell S. The rebound effect: an assessment of the evidence for economy-wide energy savings from improved energy efficiency. London: UK Energy Research Centre; 2007.
33. Walker N, Scheer J, Clancy M, Gallachoir B. Energy forecasts for Ireland to 2020. Dublin: Sustainable Energy Ireland; 2010.
34. Hinrichs-Rahlwes R. Sustainable energy policies for Europe: towards 100% renewable energy. 1st ed. Boca Ratón: CRC Press; 2013.
35. Eurostat. Electricity generated from renewable energy sources, EU-28, 2003–13 YB15. Luxembourg: Eurostat. http://ec.europa.eu/eurostat/statistics-explained/index.php?title=File:Electricity_generated_from_renewable_energy_sources,EU-28,_2003%E2%80%9313_YB15.png&oldid=238269. Accessed 4 Oct 2015.
36. Pantaleo A, Candelise C, Bauen A, Shah N. ESCO business models for biomass heating and CHP: profitability of ESCO operations in Italy and key factors assessment. Renew Sustain Energy Rev. 2014;30:237–53.
37. Traberg RL. District energy: the Danish experience. London: Royal Danish Embassy; 2008.
38. International Energy Agency. Energy technology perspectives 2015: mobilising innovation to accelerate climate action. Paris: Organisation for Economic and Co-operation and Development; 2015.
39. European Environment Agency. Progress on energy efficiency in Europe; 2015. http://www.eea.europa.eu/data-and-maps/indicators/progress-on-energy-efficiency-in-europe-2/assessment. Accessed 7 Oct 2015.
40. ENERDATA. Quantitative evaluation of explanatory factors of the lower energy efficiency performance of France for space heating compared to European benchmarks. Angers: Agence de l'Environnement et de la Maîtrise de l'Energie; 2011.
41. Eurostat. Infrastructure (electricity) annual data. Luxembourg: Eurostat. http://ec.europa.eu/eurostat/en/web/products-datasets/-/NRG_113A. Accessed 2 Oct 2015.
42. Eurostat. Combined heat and power (CHP) data 2005–2013. Luxembourg: Eurostat. http://ec.europa.eu/eurostat/web/energy/data. Accessed 2 Oct 2015.
43. Sorrell S, Mallett A, Nye S. Barriers to industrial energy efficiency: a literature review. Vienna: United Nations Industrial Development Organization; 2011.
44. Eurostat. Eurostat statistics explained: consumption of energy. Luxembourg: Eurostat. http://ec.europa.eu/eurostat/statistics-explained/index.php/Consumption_of_energy. Accessed 29 Sept 2015.
45. LEKA. New district heating countries. Vilnius: Lithuanian Energy Consultants Association; 2012.

Chapter 3
Cogeneration and District Heating Networks: Measures to Remove Institutional and Financial Barriers that Restrict Their Joint Use in the EU-28

Having that sort of market potential is probably something that's enticing them a lot more than any regulatory barriers.
—Ryan Eggers, California Energy Commission

The aim of this chapter is to identify actions that dissipate the institutional and financial barriers that are faced by those energy projects which comprise the joint use of district heating networks and cogeneration in the EU-28. From this evaluation, institutional and financial barriers were identified, which included: distinctive competence, fuel price volatility, and much of the current regulatory framework.

In order to achieve an effective removal of these barriers, and apart from generic and common actions to all the schemes, such as creating anchor loads, adopting an active marketing strategy by the local authorities, or the proposition of updating some communitarian directives on energy issues; it is also necessary to adopt those actions that respond to the casuistry of each Member State. Such actions will ultimately represent the most effective way to get a generalised implementation of energy projects that include a joint use of district heating networks and cogeneration.

3.1 Introduction

The development of the different energy infrastructures in the local energy systems, which are often disorderly, is causing growth in the number and complexity of the resulting interactions [1]. In a conventional heating system, gas is transmitted from basic transmission networks and distribution networks, located in the cities, to individual domestic boilers [2]. However, the concept is different if one considers a joint use of district heating networks and cogeneration plants. In this case, as the system expands, some externalities occur in the district heating networks [3], such that the economies of scale in the provision of heat, and a more efficient use of the fuel in individual boilers (which normally operate at partial load), can be achieved [4].

A. Colmenar-Santos et al., *District Heating and Cooling Networks in the European Union*, DOI 10.1007/978-3-319-57952-8_3

The European Commission's analysis of the evolution of the amount of cogeneration installed in the Member States between 2004 and 2008, showed an average annual growth of 0.5%, but masked a great deal of divergence in the degree of implementation of this technology. Indeed, there are only a few countries that can claim to have an effective national law resulting from the 2004/08/EC Directive on the promotion of cogeneration [5].

Even though an increasing number of local authorities are considering the joint use of cogeneration power plants with district heating networks as a technology that can be aligned with EU-28 long-term energy strategies [6] (as well as that of each one of the Member States that integrate it) and thus reduce its pollutant emissions, the fact is that, as shown in (2–4), even being one of the most CO_2-reducing technologies at a lowest cost 15, [7], there are significant barriers to both institutional [8] and market levels [9] that hamper its optimal implementation. Therefore, although it may seem that Member States' energy policies favour the development of the joint use of cogeneration power plants and district heating networks, a greater degree of investment in the necessary infrastructure for this development is not being stimulated because energy policies and regulations, far from promoting alternative solutions such as local energy infrastructures, continually reinforce a centralised energy regime [10]. In general, the various EU-28 national energy frameworks promote incremental changes and restrict some of the radical structural changes necessary to clearly [11, 12].

As an example, and as noted in most of the EU-28 Member States, efforts to achieve decarbonisation in the demand side have focused on measures such as an all-electric future [13], the development of techno-economic models that do not take into account local circumstances [13], and opportunities and incentives for energy efficiency that are based on market failures [14–17]. Such incentives tend to consider users on an individual basis, and consequently neglect the potential economies of scale that could be achieved through the combined use of district heating networks and cogeneration power plants [18]. The European Commission officially became interested in cogeneration technology in 1997, with the Community Strategy to promote Combined Heat and Power, COM(97) 514 [19]. The strategy focused on the promotion of the cogeneration power plants′ electricity generation [19].

However, this attempt by the European Commission to achieve cogeneration power plants with a lower heat to power ratio, only provoked an improvement of power plant efficiency [20, 21] and did not spread the use of district heating networks. Subsequent directives, such as Directive 2004/08/EC on the promotion of cogeneration, was not a real "positive financial stimulus" to many of the EU-28 Member States [22].

The Commission, after the observed stagnation in the diffusion of district heating since 1990 [21], developed a Directive on Energy Efficiency (the Directive 2012/27/EU). This directive explicitly recommended (in article 14) the use of district heating networks as a potential technology to meet the targets for energy efficiency, as proposed for 2020 [23]. However, in paragraph 14(7), it only alluded to "take into account" those cogeneration and district heating facilities that have a positive outcome from an economic perspective, which de facto legitimises the authorities to discretionarily choose the technology they deem as appropriate [24], and essentially, does not change the legal situation prior to that Directive.

District heating networks have particular technological and institutional characteristics (such as being a natural monopoly), which makes them a non-conventional consumer good [25]. Since a market equilibrium (where the prices are equal to marginal production costs) [26] is not achieved but a monopolistic situation, production costs will fall as production increases [27]. Consequently, it is more cost effective to have a unique infrastructure dominating the market [28]. Due to the fact that under a market failure situation private investment will not exist (or at least it will be much diminished), public investment or the development of adequate regulation is necessary [29].

In the residential sector of the EU-28, loads associated to heating equipment, ventilation, and air conditioning are those that represent a greater share of total energy consumption [30, 31], with a ratio of about 70% [32, 33]. It is expected that in the medium term, heating loads will stabilise or decrease moderately [34]. It should be noted that improved insulation [34], optimised ventilation (with heat recovery) [35], the growth of cities (which involves the creation of heat islands), and global warming [36] will lead to a decrease in the thermal load. On the other hand, the rise in population [37] and housing comfort levels [38], will contribute to the increase of the load to be heated.

From 2020, when stricter requirements to achieve zero-energy buildings are implemented [39], it is expected that the thermal heating load will decrease more rapidly (3-1) and higher primary energy savings will be obtained (Fig. 3.2).

With regard to the cooling thermal load, and although in the medium term it is expected that it will increase its importance [40], it should be noted that any estimation should be viewed with caution because of its potential subjectivity [41]. In the residential sector, it will be especially difficult to predict demand [42] because, historically, and for much of the EU, air conditioning equipment have been regarded as luxury products [43], and their acquisition and use is not exclusive to the locals [44] (Fig. 3.1).

Taking into account its weather and thermal loads, one might think that the joint use of cogeneration and district heating networks would only be profitable in Northern Europe; however, the fact that homes in these countries have better insulation has, as a consequence, specific thermal demands that are lower (Fig. 2.7). The current spread of district heating networks in the EU can be observed in Fig. 3.3. With the exception of the Nordic countries and some exceptional cases, most of these systems cover a very limited part of the city they cover.

It should remarked that for district heating networks to achieve a penetration of over 90% in a big city [46], it may take several decades (in particular, Copenhagen spent more than 40 years), its success depending greatly on the commitment of the authorities [47].

Figure B-2 (Appendix B) shows the percentage of users who have access to district heating networks for different countries belonging to the OECD (12.4% in

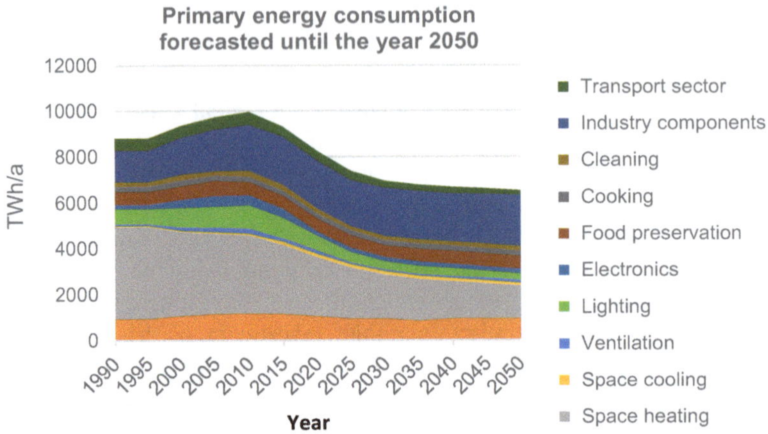

Fig. 3.1 Primary energy consumption forecasted in the EU-28 until the year 2050 (energy sector impact not shown) [40]

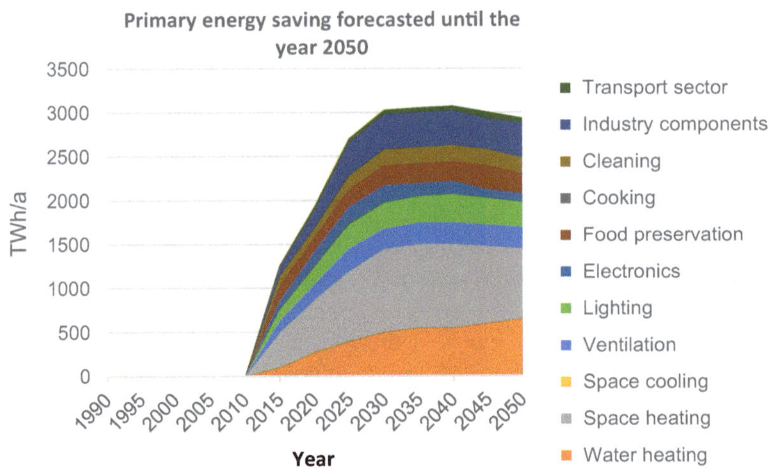

Fig. 3.2 Primary energy saving forecasted in the EU-28 until the year 2050 [40]

2011 for Member States of the European Union). Finally, in Fig. B-3, the contribution of CHP to the electricity production from different countries of the OECD is presented.

Once a first approach to the current situation of EU-28 district heating networks and cogeneration power plants has been explicated, methods to allow the work to be reproduced will be shown in section two. In section three, the institutional and financial barriers that hamper the joint use of district heating networks and cogeneration power plants, as well as actions aimed to remove the aforementioned barriers, will be presented. In section four, a discussion about the significance of the

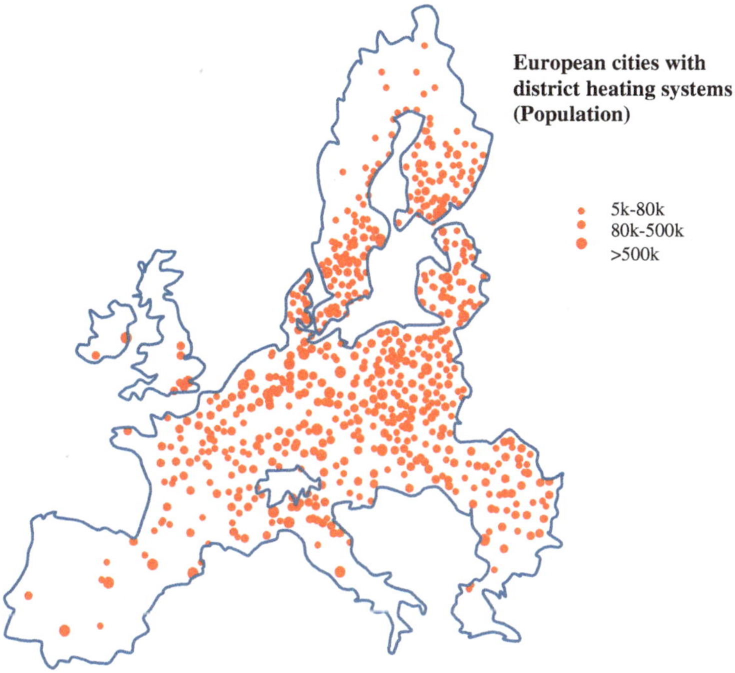

Fig. 3.3 Cities with a population of over 5000 inhabitants and district heating systems in the EU [45]

results of the work is presented. Finally, the fifth section, reserved for conclusions, presents the implications for energy policy resulting from the actions taken to face the identified barriers.

3.2 Methods

Widely discussed and accepted [15, 48], the presence of certain barriers (any element that "substantially reduces the probability of adoption of low carbon technologies") [49] are the reason for energy efficiency gap, a term which, although identified in the first instance by Blumstein [50], was coined [51] subsequently by Hirst and Brown [52] to explain the paradox of the gradual spread of seemingly profitable technologies [53]. One of the most extended definitions on energy efficiency barriers was developed by Sorrell, who considered them as "postulated mechanisms that inhibit investment in technologies that are both energy-efficient and (apparently) economically efficient" [54].

Meanwhile, market failures are those barriers that are faced by consumers (and producers) in unfettered markets and that lead to a lower market penetration than optimal from an economic point of view [48, 49, 55–57]. As Brown pointed out, for a market failure to occur, "one or more of the neoclassical economics assumptions that define an ideal market—such as perfect information and costless transactions [58]—must be violated" [15]. From a historical evaluation of the literature, this purely economic approach to explain energy efficiency barriers and its consequences on the efficiency gap was the first observed.

Despite the abundant literature, so far there is no agreement on the relative importance of each barrier [59, 60]. As to the nature of these concerns, its classification does not reveal "substantially" anything new [59]. In fact, as Weber noted, barriers are invisible, so despite being real, they cannot be empirically classified [61], given that the various existing classifications in the literature derived from the theory are diverse and relatively unstructured [56, 60]. Among the most prevalent barrier models should be the one conducted by Weber (who classifies barriers as institutional, economical, organizational, and behavioural) [621] that has been adopted by other researchers such as Rodhin [62] or Sorrell [54], Thollander (who frames barriers in technical systems and technological and socio-technological regimes) [57] and Hirst, who divides barriers into structural and behavioural categories [52]. However, the taxonomy adopted in this book is an adaptation of the one proposed by Chai [59], grouping barriers into the following categories: behavioural, market failures, physical constraints, and institutional and financial barriers, the latter two being the only barriers addressed in this chapter.

Recently, a different approach from researchers to explain energy efficiency barriers has been gaining acceptance (conducting interviews with and providing questionnaires to experts in order to identify, classify, and discuss the barriers according to their nature [59] and thus propose measures to overcome them). While it is true that, as argued by Chai and Yeo, results of these interviews are inherently "contingent" [59], equally veritable is the fact that the barriers identified by the experts consulted in the writing of this chapter are quite similar to each other (however, it is true that there are significant variations in the relative importance each of the respondents places on each barrier).

In order to identify the barriers to the implementation of district heating networks and cogeneration in the EU-28 and due to their nature, an extensive review of (updated) academic and grey literature on energy barriers was first conducted, using the preliminary results to proceed with contacting experts in this field. From their experiences, information was obtained on what the barriers are, how they can be eliminated/mitigated and if current measures proposed by the governments of their respective countries are adequate. About 150 publications were reviewed for this chapter, and despite being a substantial and representative sample of the barriers identified in the literature, this number is not absolute.

It should be noted that in this chapter, barriers and market failures have been differentiated (some authors, such as Geller and Attali [63], consider this differentiation useful or even compulsory [49] as Gillingham and Sweeney point out) and that the evaluation of those institutional and financial barriers (and not market failures) that hamper the joint use of district heating networks and cogeneration in the EU-28 (as well as the proposal of corrective measures) is the only scope of this chapter. The reason for this approach is the fact that market failures are well known and understood [49], while institutional and financial barriers (in sensu stricto), despite having a "particular relevance to the implementation of energy efficient technologies", have been subject to much less attention and lack adequate empirical support [49].

Historically, efforts have focused on getting a more efficient energy market, assuming a purely economic (and not technological) perspective [54, 60] aimed at removing market failures [48]. Neoclassical economic theory (which, as Sorrell points out, while useful and dominant in the literature, is insufficient to explain energy-efficiency barriers) [54] suggests that in the absence of market failures, there should not be a big difference between optimal and real energy efficiency, being as the regulations in this sector are particularly "undesirable" [60, 64].

However, the technological point of view [54] (with a higher potential than the economic one) as defined by Jaffe and Stavins [53], is adopted in this book as its aim is finding the socially optimum energy-efficiency level: a level that can only be achieved through measures that eliminate or at least mitigate not only market failures but also barriers [48]. In this sense, the view proposed in this book (the technological one) is in line with the approach favoured by the EU [65]. Certainly, public intervention in energy efficiency is both an empirical and ideological issue [54].

The barriers identified by Chai [59], Sorrell [54], and Brown [15] were transferred to 19 experts on cogeneration and district heating networks from nine different countries (eight of them European), who contributed either through interviews, e-mail correspondence, and/or surveys (where the experts identified the barriers and the relative importance of each of them) to give validity and reliability to the identified barriers in the literature. It should be noted that, unlike uncertainty or lack of attention from institutions to energy issues (which themselves are barriers in the strict sense), some of the proposed barriers by the experts consulted (such as asymmetric information and imperfect information), are themselves market failures [53, 54, 59] and, consequently, will not be expanded in this chapter.

3.3 Institutional and Financial Barriers Identified

3.3.1 Distinctive Competence and Business Model Effect

If the EU-28 Member States do not create a regulatory framework aimed at converting energy efficiency into part of the utilities' distinctive competence [63], then in spite of the implementation of cogeneration power plants and district heating networks being technically and economically feasible, the schemes cannot progress due to the fact that investment in cogeneration is normally less attractive than other projects that frame, focus and do not conflict with its distinctive competence [64]. These projects also offer a better rate of return, a lower intrinsic risk, and can be implemented more easily due to the fact that they do not have to overcome the barriers associated with cogeneration projects [65–67].

It is clear (and experience in other sectors with natural monopolistic characteristics has shown) that utilities will not embark on entrepreneurial or innovative activities without an incentive to do so [68], and this incentive should undoubtedly come from the regulator [69].

Empirically, it has been found that efforts to encourage focal organisations (those that serve as a reference point), such as distribution network operators, to act as institutional entrepreneurs, has been problematic [70]. This is due to the fact that the entrepreneurship is contrary to the competitive logic that has been established in virtually all of the most developed EU countries. Innovation and change are by no means among the priorities of focal organisations that benefit from a privileged status quo [71, 72].

3.3.2 Fuel and Electricity Price Volatility

The relationship between the prices of gas and electricity is known as *spark spread* [73]. The use of cogeneration will be more attractive the lower the price of gas is relative to electricity, i.e., when the *spark spread* is higher [74], and this factor is the one with the highest impact on both the internal rate of return and on the level of risk associated with a cogeneration project [75]. In recent years, the rise and volatility of gas prices relative to electricity (low spark spread) has been a major deterrent to the expansion of cogeneration, and it is expected this trend will continue in the near future [76].

Despite having the option to enter into long-term contracts in order to deal with the risk associated with the high volatility of gas prices in the market, the fact is that it is precisely this volatility which makes access to this contract by large users to be priced considerably higher [77]. Consequently, the *spark spread* between the prices of gas and electricity shrinks, lowering the financial attractiveness of cogeneration [78] due to the fact that the volatility experienced by the price of gas is normally not exactly reproduced in the price of electricity [79]. This additional risk will increase the required rate of return for embarking on a project that involves the implementation of a cogeneration system with a district heating network [80].

In this section, it should be noted that, unlike other EU-28 members, and in order to reduce the uncertainty about the price of a particular energy resource, producers in countries with a long tradition in cogeneration, such as Sweden, can normally choose discretionally between at least two fuels, enabling flexibility to a sector of economically feasible cogeneration in Sweden [81].

3.3.3 Other Reasons

The institutional and legal framework for district heating networks mainly belongs to the national level of each Member State. One of the very few international directives that affects the competitiveness of the district heating networks, the European Union Emissions Trading System [82, 83] and the Directive on Energy Efficiency (the Directive 2012/27/EU) that, as indicated above, and despite explicit recommendations for the use of district heating to meet the EU-28 energy goals set by 2020, does not legally oblige its members to implement this technology even if it is economically feasible. So, we can say that as of today, there is no legislation at a supranational level that specifically ensures an expansion of district heating networks. Apart from the effect of the distinctive competence, the business model, and fuel and electricity price volatility, the following barriers have been identified that also hinder the development of the district heating networks:

- Long-term investment: District heating involves a long-term commitment [84], comparable to other public works projects of relevance. This makes it unattractive to those energy markets that have already been privatised and opened to competition, since they prefer projects with shorter payback periods [85]. This fact, coupled with the higher risk involved in the implementation of district heating in comparison to other more conventional technologies, makes the required cost of capital greater [86, 87].
- Regulatory framework: The influence of a different regulatory framework for each Member State and, in many cases, a heterogeneous approach, makes the EU-28 district heating situation completely unharmonised [88], and hinders the entry of those market players who have previously participated in projects of the same nature in other Member States.
- Regulation and energy price distortion: Most Eastern European countries enforce a heat price regulation to protect the most disadvantaged part of the population (mainly because welfare programs have not been introduced at a national or local level) [89]. This causes not only an energy market with distorted prices [90] but, indirectly, prevents a proper maintenance and expansion of district heating networks, as many private investors are prevented from entering the district heating sector [91].
- Energy market liberalisation: Different models adopted to liberalise the energy market in the different Member States have caused a number of difficulties for market players. In particular, to the lack of integration of the energy market [92], joins the fact that, paradoxically, market liberalisation has caused an atomisation of the energy industry into a small group of large companies [93–95] that are not interested in the development of cogeneration [96].
- Priorities, experience and electioneering of local authorities: Unlike health and education, local authorities in the EU-28, in addition to not having experience in the matter in question, do not consider energy as a priority [97], and many of the actions carried out in this area, if done, are at least "non-transparent" [98]. On the other hand, because local authorities are knowledgeable that, although beneficial to society as a whole, electorally it would be dangerous to go into debt for energy projects that are not essentially a claim or object of open discussion within the general public, they will in most cases choose not to embark on such projects [99].

Table 3.1 Most significant EU-28 institutional and financial barriers to the take off district heating networks and cogeneration

Identified barriers	Reasons
Distinctive competence and business model	Lack of a policy framework that aims to make energy efficiency part of the distinctive competence of the utilities Generally, focal energy organisations benefiting from a privileged status quo lack a desire for innovation and change
Fuel and electricity price volatility	High gas price volatility Low spark-spread in recent years
Long-term investment	Payback period longer than other energy projects Risk of implementation higher than other, more conventional technologies
Regulatory framework	Lack of harmonisation of the rules relating to district heating for the EU-28 Member States
Regulation and distortion of energy prices	Lack of social assistance programs at the national or local level in some of the Eastern European countries belonging to the EU-28
Energy market liberalisation	Lack of integration of the energy market Atomisation of the energy industry
Priorities, experience and electioneering of the local authorities	Generally, local authorities do not consider energy to be a priority Potential electoral danger for the debts that would result from the execution of this type of energy projects

Having identified the main institutional and financial barriers that hamper a further development of district heating networks, it [sic] is shown in Table 3.1 by way of the summation set forth in this section.

3.4 Measures Aimed at Eliminating Existing Institutional and Financial Barriers

Throughout this section, a number of identified institutional and financial barriers related to the introduction of district heating networks and cogeneration in the EU-28 have been presented; Table 3.2 shows proposed measures to combat them.

Each of the mechanisms proposed below shall be understood as designed to be flexible enough to meet changing requirements and to prevent redundant grants/financial support.

Table 3.2 Measures proposed to address the institutional and financial barriers that hamper the development of district heating networks and cogeneration in the EU-28

Identified barrier	Measures proposed
Distinctive competence and business model	Complete elimination of risk infrastructure imposing a discount rate of 3.5% Forcing the national energy regulator of each Member State to involve focal organisations on energy infrastructures of this kind
Fuel and electricity price volatility	To offer a guaranteed minimum price to energy providers Change tariffs that attend to an alternative fuel cost index Maintain infrastructure profitability within a given threshold range, with an annual review of gas prices, of the spark spread and of the financial support level
Long-term investment	Provision of support for the electricity produced from high-efficiency cogeneration power plants Modification of payment terms of district heating networks providers Administrative intervention by local authorities, so that standardised contract terms are provided Obligation to the newly built power plants to have a minimum efficiency of at least 70%
Regulatory framework	Creating anchor-loads representing a load of about 80% of the total capacity of the district heating network Updating the Directive 2009/72/EC in order to explicitly allow to provide long-term contracts to those suppliers of technologies that contribute to compliance with environmental obligations of the EU-28 Member States Creating an organisation belonging to the central government that would develop the political framework and would act as a leader in the development of district heat networks
Regulation and distortion of energy prices	Introduction of assistance programs at a national or at a local level in those Member States where they are not present Implementation of tariff differentiation or other form of compensation for those cases where district heating networks are more expensive for the consumer than the conventional alternative
Energy market liberalisation	Implementation of a pure carbon model
Priorities, experience and electioneering of the local authorities	Make adjustments to planning and building regulations Implement a strategy of institutional and active marketing by local authorities with programs of promotion and dissemination of the technology Conduct demonstration projects of cogeneration power plants and district heating networks Perform an institutional innovation through the creation of a public company only in those Member States that do not have favourable market conditions and where the above measures have proven ineffective

3.5 Discussion

The Community guidelines on State aid for environmental protection (Communication 2008/C 82/01) recognise that, in certain circumstances and with the objective to promote broader environmental policy goals, State aid may be necessary [100]. These guidelines, which have been adopted in close collaboration with Member States, intend to serve as a tool for the promotion of measures to protect the environment while preventing unjustified State aid [100], and are fully consistent with the European Directive 2004/8/EC on the promotion of cogeneration [100]. Therefore, the Commission's position is that State aid is acceptable under certain conditions if it serves to promote the use of renewable energy sources as well as the combined heat and power production [100].

However, as discussed in previous sections, the evolution of the implementation of one of the technologies that reduces emissions at a lower cost (such as district heating networks), has been in the last 25 years truly daunting for the vast majority of Member States. If the Member States really have an energy and environmental commitment, they should seek alternatives to the actions already taken. These measures should have an ultimate goal to address the barriers identified in the previous section. Even those that may be generically assumed by all Member States should be sought individually, depending on the existing level of maturity and the implementation of district heating networks.

Currently, district heating networks contribute about 10% of the total EU heat supply [101]. A spontaneous change of this circumstance without the required political, socio-cultural, and market environments is unlikely [102–104]. Nevertheless, if some requirements were met, it is feasible that district heating will become a major player in the European energy sector. In order for this to happen, the risks and costs associated with this type of project must be mitigated [105].

In previous sections, the reasons limiting the implementation of the joint use of CHP with district heating networks and a set of proposed measures were presented; this section will discuss the significance of the results of this chapter.

Measures to mitigate the barrier of distinctive competence and business model: In the previous section, the reasons why focal energy organisations do not consider district heating networks or cogeneration attractive were presented. Among them, the intrinsic risk of these infrastructures and the payback period were highlighted. The national energy regulators of each Member State were proposed to bear the responsibility of encouraging the participation of these organisations.

In general, since the different privatisations across the energy sector in the EU-28 Member States, investment in energy infrastructure has been subject to market forces [106] that do not encourage to invest and protect critical infrastructures [107]. We propose that, in order to mitigate the reluctance of the aforementioned focal energy organisations to participate in the launch of cogeneration and district heating networks, the risk should be completely eliminated, imposing a discount rate (typically used in evaluating projects where, somehow, the public sector is involved and represents the threshold project profitability) of 3.5% [108–110], forcing the national energy regulator of each Member State into the participation of these focal organisations on infrastructures of a similar nature.

It should be clear that this action should only be taken in those energy markets where district heating networks technology is not mature enough. The objective would be for the action to not only be developed and implemented in a particular Member State but also to become part of the distinctive competence of focal energy organisations.

Measures to mitigate the barrier of fuel and electricity price volatility: To promote private investment, its financial profitability should be attractive [111] as constant as possible and with certainty on how the government will behave on tariff adjustments [112]. Although a guaranteed floor price to energy providers has been proposed, investors should accept that energy prices are unpredictable and volatile [113], so it is necessary for them to tolerate this fact.

Regardless of the size of the schemes and due to the characteristics of effective monopoly that take place [25, 111], the adoption of an appropriate form of regulation for the sake of modifying tariffs attending to an alternative fuel cost index has been proposed, thereby maintaining competition between different fuels and technologies.

Without becoming a barrier to investment itself, but accepting that it adds considerable complexity to the administration of the scheme and the evaluation undertaken by investors, it has been proposed as a measure to mitigate the volatility of fuel prices and electricity, maintaining the profitability of the developed infrastructure within a (modest) limit given, conducting an annual review of gas prices, spark-spread, and level of financial support.

Measures to eliminate or mitigate the barrier of long-term investment: The most immediate measure to eliminate or mitigate the barrier of long-term investment is providing financial support that would raise the internal rate of return, decrease the risk of investment, and stimulate the spread of district heating networks, thus acting as a catalyst for its implementation [113]. It has been argued that as the disbursement taken to implement district heating networks is far outweighed by the great benefits they have for society [114], EU-28 governments should support cogeneration and district heating networks in order to avoid the market failure resulting from externalities that arise from the conventional production of heat and electricity. Although a practical solution might be providing low-interest loans or offering investment subsidies [115], it should be noted that, if done, any grant scheme should be limited in each Member State to a set number of projects, designed to illustrate both the feasibility of implementing district heating networks as the catalysts for reducing its cost and the formation of a local supply chain.

Therefore, the grant scheme will vary greatly, depending on the maturity of each Member State in terms of district heating networks. Historically, there have been problems with direct subsidies to infrastructure installation, both in terms of what they are used for and their administration [116]. Therefore, this is not considered to be the most advisable action to take. Providing subsidies that cover infrastructure costs might mean that they are "open to abuse" [117, 118].

The provision of aid for electricity produced from high-efficiency cogeneration power plants, not only would reduce significantly the payback of the investment and would provide a guaranteed income over the life of the cogeneration power plant and the district heating networks, but also would mitigate the perceived risks, so we see this measure as one of the ones that should be taken to mitigate the barrier of long-term investment.

Although the level of financial aid can vary to reflect market conditions, it is important to bear in mind the need for maintaining it for projects that have already been committed under this scheme. Although financial aid would be different for each EU-28 country, successful results obtained in similar schemes in Denmark [119], as well as studies conducted in less-developed countries (such as Croatia) in terms of district heating networks and cogeneration [22], indicate, by way of guidance, that they should range from 0.005 €/kWh to 0.01 €/kWh per unit of electricity produced from high-efficiency cogeneration.

Theoretically, with competitive market incentives and an appropriate regime, private investment on its own should be attracted and it could be used to promote efficient use of district heating [120] and cogeneration. However, an additional element is necessary, one that reduces the risk of the infrastructure, such as the modification of payment terms required by district heating networks suppliers, giving them the opportunity to receive guaranteed payments (similar to those provided in contracts where the public-private partnership occurs) [121] and we propose this as a measure to eliminate or mitigate the long-term investment barrier.

It should be noted that, in general, schemes that have been successful in Europe have had a high degree of public sector involvement [122, 123], and normally, local authorities have had the ultimate responsibility for the final push of the development of district heating networks [124]. In response to this experience and the need to eliminate the barrier of long-term investment, a high involvement of local authorities has been proposed, whose functions would be (besides enacting the spread of district heating networks) to carry out an administrative intervention to provide standardised contract terms in order to attract users using heating technologies other than district heating networks.

Finally, and as far as the construction of new power plants is concerned, the requirement to have a minimum efficiency of at least 70% was proposed. This would make cogeneration a low-cost alternative, create a strong incentive for them to find sustainable heat markets, and remove long-term investment barriers.

Measures to eliminate or mitigate the barrier of the regulatory framework: To reduce the risk associated with these schemes, in particular those related to oversizing a potentially obsolete system, the creation of anchor loads that represent roughly four out of five clients through long-term contracts has been proposed.

In most countries where district heating networks have positively thrived, providers offer contracts of up to 20 years to ensure income [125, 126]. However, this seems to go against the energy policy of many Member States, since it could lead to unintended consequences for the current competition of energy markets (it is likely a competition distortion between district heating networks and other technologies) as well as for the protection of heat consumers. If they are subjected to long-term contracts, there might be problems related to the abuse of dominance [127].

In the Directive 2009/72/EC of the European Parliament and of the Council concerning common rules for the internal market in electricity, it is stated that "Member States shall ensure that the eligible customer is in fact able easily to switch to a new supplier" [128], which obviously creates a conflict between the legal imperative to give the user the power to change suppliers and the requirement of having long-term contracts to reduce the risk of investing in infrastructures of this type. Due to the abovementioned conflict and to the approaches and different interpretations among Member States, an updating of the Directive 2009/72/EC is proposed to remove regulatory barriers in order to explicitly allow providers of those technologies that serve to fulfill the environmental obligations incurred by Member States for the year 2020 long-term contracts. This will ensure not only the necessary income f or suppliers, but also a fulfillment of such obligations by the Member States.

As set forth above, local authorities should be the ones to give the final impetus to these kinds of infrastructures, depending the success of the scheme on its performance. However, without the involvement of the central government in establishing the political framework in which local authorities operate, imposing obligations and the provision of the appropriate authority thereto will be very difficult in terms of district heating networks [129–131].

Therefore, the creation of a governmental institution or assimilated body which would undertake these duties and become the main champion for the development of district heating networks is proposed as a measure to eliminate (or mitigate) the barrier of the regulatory framework. In those Member States where technology is not mature enough, it is proposed to add some measures to strengthen the regional supply chain related to this technology.

Measures to eliminate or mitigate the barrier of the regulation and distortion of energy prices: In order to eliminate or mitigate this barrier, the introduction of social assistance programs at a national or local level is proposed in order to ensure unnecessary the implementation of heat prices that protect the most vulnerable part of the population, as they would ultimately impede the proper maintenance and expansion of district heating networks.

In spite of the aforementioned measure to remove the barrier of the regulation and distortion of energy prices, it should be noted that the expansion of district heating networks may raise issues regarding cross-subsidies and consumer protection [128, 132–135]. For example, in Denmark 8% of district heating customers pay more than if they used an individual natural gas furnace [136]. To minimise this circumstance, the implementation of tariff differentiation or a form of compensation that addresses this situation is proposed.

Measures to mitigate the barrier of energy market liberalisation: In order to mitigate this barrier, the different models adopted to liberalise the energy market of the Member States are proposed to be replaced by a new carbon market model. The new model, in which the shadow price of carbon would be fully assigned to the combustion of fossil fuels, would fully remunerate the emissions released into the atmosphere due to more efficient schemes through the carbon price. This measure, coupled with the ones outlined above to mitigate the barrier of distinctive competence and business model, according to which national energy regulators from each Member State would require the participation of focal energy organisations on district heating networks and cogeneration projects, would provoke the little interest shown by them (even further intensified after the progressive atomisation of the energy sector following the sector liberalisation) regarding these technologies to be reversed.

Measures to eliminate or mitigate the barrier of the priorities, experience, and electioneering of local authorities: In countries where district heating networks are widely deployed and in those that have managed to eliminate the barriers associated with the priorities, experience, and electioneering of local authorities, the planning carried out by the municipalities has been vital in its development. Although, in the first instance, potential users of district heating networks will be public buildings, shopping centers and hotels [70, 137]; the attractiveness is even higher for electrically heated dwellings. We propose making adjustments to the regulations of planning and construction, showing builders the potential benefits of its use, so that they will at least consider district heating as an option (district heating networks have the advantage that, for these particular types of buildings, they are not competing with any network already implanted with large sunk costs, with the downside of the better building construction efficiencies at present).

Following the example of the political lines carried out by Denmark [138], it was pondered as a measure to obligate the connection to the district heating scheme in certain areas and circumstances. Even being aware that this action would not only involve a much faster method of getting the desired results than depending on a spontaneous evolution of the private sector toward it, but a total cost lowering and the consumer favouring [139], this measure could not be justified because it would not fit well with the current choice for freedom and market competition existing in most of the EU Member States; therefore, it will not be proposed. There are other forms for the promotion of district heating networks that are less imposing than those carried out in Denmark, and that would adapt better to the aforementioned energetic freedom. Instead of forcing the connection, what we propose is that the local authority employ any form of institutional and active marketing.

It should be remarked that, if connection is voluntary, then contractual terms should cover a least the duration of the contract, conditions for extending or terminating it, service quality, and arrangements for compensation [140, 141]. However, and as obtaining a high degree of connections at the private level is at least complicated, obliging public authorities and public housing providers to connect to the district heating network as well as explaining to new buildings constructors its benefits is proposed in order to counteract an individual that is, per se, change and risk-reluctant and know beforehand its profit resulting from using this technology is not significant [142–145].

Furthermore, carrying out feasibility studies of technologies/innovative applications, demonstration projects, and programs for the promotion and dissemination (they will work well as long as they do so over a long period of time and are supported by favourable market conditions) of cogeneration power plants and district heating networks is proposed in order to mitigate the barriers associated with the priorities and electioneering of local authorities.

As noted, these measures will be ineffective without favourable market conditions. In those Member States where this situation occurs and the technology is not seated, conducting an institutional innovation through the creation of a public company for the development of district heating networks with cogeneration facilities is proposed as a way to raise awareness about its existence. This measure is based on the fact that if a given government (either national or local) is determined to create companies that combine all the necessary skills for the implementation of district heating networks with cogeneration facilities, then transaction costs of the technology will decrease significantly. The organisation will, in all likelihood, be able to finance long-term access to capital (which is, on the other side, higher), involving, in the first instance, the easiest private clients (which, in this case, would be the existing industries and power plants). The initial phase would raise the awareness of the existence of this technology and its associated benefits. This would result in the short-term removal of the detected electoral barrier, as potential debts of such projects would be understood by the general public. In the medium-term, more complicated private clients would be attracted.

3.6 Conclusions and Policy Implications

The existing institutional and financial barriers to the implementation of district heating networks with cogeneration power plants in the EU-28 were identified. It was found that unless a substantial change in the market, or in the regulation concerning the use of the above technology, occurs, a definitive take off thereof would be difficult. The most recent EU Directives (among which Directive 2012/27/EU on energy efficiency stands out), despite being willful, are insufficient to achieve a spread of this technology according to its potential.

The reduction of energy poverty has been determined as imperative in order to achieve the targets for energy efficiency proposed within the EU-28 by 2020. Furthermore, the development of energy policies that create a symbiotic relationship between the public and the private; a more accurate assessment of environmental externalities; and the removal, in line with our proposals, of the current institutional and financial barriers for the joint use of district heating and cogeneration; are also required to achieve the EU-28 2020 targets.

For his part, it has been determined that, because currently the joint use of cogeneration and district heating networks is in most of the Member States at an early stage, the most effective way to develop sector institutions, technical regulations, and legal and contractual provisions, is that of conducting an individualised search for measures. That, mainly according to their degree of maturity and existing implementation in each Member State, evolve and lead as the system expands to an appropriate regulatory regime. Based on what is stated here, that would eliminate the institutional and financial barriers that those energy projects involving the joint use of cogeneration and district heating networks face.

References

1. Rad FD. On sustainability in local energy planning. Lund: University of Lund; 2011.
2. Agency London Climate Change. Moving London towards a sustainable low-carbon city: an implementation strategy. London: London Climate Change Strategy; 2007.
3. Corvellec H, Zapata-Campos MJ, Zapata P. Infrastructures, lock-in, and sustainable urban development: the case of waste incineration in the Göteborg metropolitan areas. J Clean Prod. 2013;50:32–9.
4. Kavalov B, Peteves SD. Bioheat applications in the European Union: an analysis and perspective for 2010. Luxembourg: European Commission; 2004.
5. CODE. Growing cogeneration in Europe: D6,1 proposal for a European cogeneration roadmap. Brussels: Cogeneration Observatory and Dissemination Europe; 2011.
6. Greater London Authority. District heating manual for London. London: Greater London Authority; 2013.
7. Connolly D, Lund H, Mathiesen BV, Werner S, Möller B, Persson U, et al. Heat roadmap Europe: combining district heating with heat savings to decarbonise the EU energy system. Energy Policy. 2014;65:475–89.
8. Commission European. Action plan for energy efficiency: realising the potential. Brussels: European Commission; 2006.
9. Verbruggen A. The merit of cogeneration: measuring and rewarding performance. Energy Policy. 2008;36(8):3069–76.
10. Rydin Y, Turcu C, Austin P. Planning and the challenge of decentralised energy: a co-evolution perspective. Stockholm: 2011 Nordic Environmental Social Science Conference; 2011.
11. Mitchell C. The political economy of sustainable energy. Basingstoke: Palgrave Macmillan; 2010.
12. Lockwood M. System change in a regulatory state paradigm: the "Smart" grid in the UK. Bordeaux: European Consortium for Political Research General Conference; 2013.
13. Speirs J, Gross R, Deshmukh S, Heptonstall P, Munuera L, Leach M, et al. Building a roadmap for heat 2050 scenarios and heat delivery in the UK. London: Combined Heat and Power Association; 2010.
14. Golove WH, Eto JH. Market barriers to energy efficiency: a critical reappraisal of the rationale of public policies to promote energy efficiency. Berkeley: Lawrence Berkeley National Laboratory; 1996.
15. Brown MA. Market failures and barriers as a basis for clean energy policies. Energy Policy. 2001;29(14):1197–207.
16. Hynek D, Levy M, Smith B. Follow the Money: overcoming the split incentive for effective energy efficiency program design in multi-family buildings. Pacific Grove: ACEEE Summer Study on Energy Efficiency in Buildings; 2012.
17. Department of Energy & Climate Change. The energy efficiency strategy: the energy efficiency opportunity in the UK. London: Department of Energy & Climate Change; 2012.
18. Piel E. Preliminary contribution to the Commission's Public consultation on the review of state aid guidelines for environmental protection. Brussels: Euroheat & Power; 2012.
19. Commission European. A Community strategy to promote combined heat and power (CHP) and to dismantle barriers to its development. Brussels: European Commission; 1997.

20. Grohnheit PE. Energy policy responses to the climate change challenge: the consistency of European CHP, renewables and energy efficiency policies. Roskilde: Risø National Laboratory; 1999.
21. Zmijewski K. Current challenges in completing the internal energy market. Brussels: EPP Group Hearing on the Internal Energy Market; 2013.
22. Lončar D, Ridjan I. Medium term development prospects of cogeneration district heating systems in transition country—Croatian case. Energy. 2012;48(1):32–9.
23. Commission European. Directive 2012/27/EU of the European Parliament and of the Council of 25 October 2012 on energy efficiency, amending Directives 2009/125/EC and 2010/30/EU and repealing Directives 2004/8/EC and 2006/32/EC. Brussels: Eurpean Commission; 2012.
24. Ends Europe. Draft working paper on article 14 of the energy efficiency directive (2012/27/EU) (Version 20/11/2012 for the meeting on 04/12/2012). London: Ends Europe; 2012.
25. Söderholm P, Wårell L. Market opening and third party access in district heating networks. Energy Policy. 2011;39(2):742–52.
26. Osborne MJ. Short run competitive equilibrium in an economy with production. Toronto: University of Toronto; 1997. http://www.economics.utoronto.ca/osborne/2x3/tutorial/LRCE.HTM. Accessed 15 Aug 2014.
27. The Linux Information Project. Natural monopoly definition. Bellingham: The Linux Information Project; 2005. http://www.linfo.org/natural_monopoly.html. Accessed 15 Aug 2014.
28. ESMAP. Increasing the energy efficiency of heating systems in Central and Eastern Europe and the former Soviet Union. Washington: The World Bank; 2000.
29. Killick T, White H, Savane MA, Kayizzi-Mugerwa S. African poverty at the millennium: causes, complexities, and challenges. Washington: World Bank; 2001.
30. Nguyen TA, Aiello M. Energy intelligent buildings based on user activity: a survey. Energy Build. 2013;56:244–57.
31. Lu J, Sookoor T, Srinivasan V, Gao G, Holben B, Stankovic J, et al. The smart thermostat: using occupancy sensors to save energy in homes. In: Zurich: 8th ACM Conference on Embedded Networked Sensor Systems; 2010.
32. Economidou M, Atanasiu B, Despret C, Maio J, Nolte I, Rapt O. Europe's buildings under the microscope: a country-by-country review of the energy performance of buildings. Brussels: Buildings Performance Institute Europe; 2011.
33. Gasparella A, Baratieri M. Analysis of the energy performance of the building envelope and HVAC systems. Bolzano: Free University of Bozen-Bolzano; 2014. https://www.unibz.it/en/sciencetechnology/welcome/AnalysisoftheenergyperformancofthebuildingenvelopeandHVACsystems.html. Accessed 20 Aug 2014.
34. Union European. EU Energy, transport and GHG emissions trends to 2050: reference scenario 2013. Luxembourg: European Union; 2014.
35. NIBE Industrier AB. Market trends. Markaryd: NIBE Industrier AB (publ); 2014. http://www.nibe.com/The-Group/NIBE-operations/We-have-created-a-leading-European-supplier-of-sustainable-energy-solutions/The-year-in-brief/. Accessed 20 Aug 2014.
36. McCarthy M, Best MJ, Betts RA. Climate change in cities due to global warming and urban effects. Geophys Res Lett. 2010;37(9):L0905.doi:10.1029/2010GL042845.
37. European Commission. EU Employment and social situation. Quaterly review: March 2013. Special supplement on demographic trends. Luxembourg: European Commission; 2013.
38. DG for Energy and Transport. Labelling and other measures for heating systems in dwellings: Final technical report. Brussels: European Commission; 2002.
39. Commission European. Progress by Member States towards Nearly Zero-Energy Buildings. Brussles: European Commission; 2013.
40. Kemma R. Ecodesign impact accounting: Part 1-Status Nov. 2013. Brussels: European Commission; 2014.

41. ASHRAE. Fundamentals. ASHRAE handbook. Atlanta: American Society of Heating, Refrigerating and Air-Conditioning Engineers; 2005. p. 2005.
42. Heschong Mahone Group. The affordable housing energy efficiency handbook—A guide on how to incorporate energy efficiency into affordable housing: new construction and rehabilitation projects. Gold River: Heschong Mahone Group; 2008.
43. Town of East Gwillimbury. Community energy plan: status update. East Gwillimbury: Town of East Gwillimbury; 2009.
44. Carbontour. Strategic planning towards carbon neutrality in tourism accommodation sector: Recording and assessment of the existing situation regarding tourism development in North Aegean, Greece and Cyprus—selection of the eligible accommodation units for participating in the Project. Zografou: National Technical University Of Athens; 2011.
45. Universitet Aalborg. Heat roadmap Europe 2050. Aalborg: Aalborg Universitet; 2013.
46. DHC + Technology Platform. District heating & cooling: a vision towards 2020–2030–2050. Brussels: DHC + Technology Platform; 2009.
47. Thornton R. Copenhagen's district heating system: recycling waste heat reduces carbon emissions and delivers energy security. Westborough: International District Energy Association; 2009.
48. Backlund S, Thollander P, Ottosson M, Palm J. Extending the energy efficiency gap. Energy Policy. 2012;51:392–6.
49. Gillingham K, Sweeney J. Barriers to implementing low carbon technologies. New Haven: University of Yale. http://www.yale.edu/gillingham/GillinghamSweeney_CCE12_BarriersLowCarbonTechs.pdf. Accessed 15 Aug 2014.
50. Blumstein C, Krieg B, Schipper L, York C. Overcoming social and institutional barriers to energy conservation. Energy Policy. 1980;29(14):1197–207.
51. Rosenow J. Home energy efficiency policy in Germany and the UK. Oxford: Earthscan; 2013.
52. Hirst E, Brown MA. Closing the efficiency gap: barriers to the efficient use of energy. Resour Conserv Recycl. 1990;3(4):267–81.
53. Jaffe AB, Stavins RN. The energy-efficiency gap: what does it mean? Energy Policy. 1994;22(10):804–10.
54. Sorrell S, Schleich J, Scott S, O'Malley E, Trace F, Boede U, et al. Barriers to energy efficiency in public and private organisations. Brighton: Science Policy Research Unit—University of Sussex; 2000.
55. Vaidyanathan S, Nadel S, Amann J, Bell CJ, Chittum A, Farley K, et al. Executive summary: overcoming market barriers and using market forces to advance energy efficiency. Washington: American Council for an Energy-Efficient Economy; 2013.
56. Dunstan C, Daly J. Institutional barriers to intelligent grid: a discussion paper. Sydney: University of Technology Sydney; 2008.
57. Thollander P, Palm J, Rohdin P. Categorizing barriers to energy efficiency—an interdisciplinary perspective. In: Palm JE, editor. energy efficiency. Rijeka: Intech; 2010. p. 49–62.
58. Hayek F. Critical assessments of contemporary economists. New York: Routledge; 1991.
59. Chai KH, Yeo C. Overcoming energy efficiency barriers through systems approach—a conceptual framework. Energy Policy. 2012;46:460–72.
60. Sorrell S, Mallett A, Nye S. Barriers to industrial energy efficiency: a literature review. Vienna: United Nations Industrial Development Organization; 2011.
61. Weber L. Some reflections on barriers to the efficient use of energy. Energy Policy. 1997;25(10):833–5.
62. Rodhin P, Thollander P, Solding P. Barriers to and drivers for energy efficiency in the Swedish foundry industry. Energy Policy. 2006;35(1):672–7.
63. Geller H, Attali S. The experience with energy efficiency policies and programmes in IEA countries: learning from the critics. Paris: International Energy Agency; 2005.
64. Blumstein C, Taylor M. Rethinking the energy-efficiency gap: producers, intermediaries, and innovation. Berkeley: Energy Institute at Haas; 2013.

65. Costa MT, García-Quevedo J, Segarra A. Energy efficiency determinants: an empirical analysis of Spanish innovative firms. Barcelona: Institut d'Economia de Barcelona; 2014.
66. Public Service Commission. Comments of the joint utilities. New York: New York State; 2013.
67. Division of Ratepayer Advocates. 2011 Annual Report. San Francisco: California Public Utilities Commission; 2012.
68. Global Environment Facility. Removing barriers to the increased use of biomass as an energy source. Washington: United Nations Development Programme; 2001.
69. McNaught C. A study into the recovery of heat from power generation in Scotland. Glengamock: AEA; 2011.
70. King M. Community energy: planning, development and delivery. Westborough: International District Energy Association; 2012.
71. Bauknecht D, Leprich U, Späth P, Skytte K, Esnault B. DG-Grid: regulating innovation & innovating regulation. Brussels: Intelligent Energy; 2007.
72. Hawkey D, Webb J. Multi-level governance of socio-technical innovation: the case of district heating in the UK. Copenhagen: Copenhagen Business School; 2012.
73. Schofield N. Commodity derivatives: markets and applications. Chichester: Wiley; 2007.
74. Shahidepour M, Yamin H, Li Z. Market operations in electric power systems: forecasting, scheduling, and risk management. New York: Wiley; 2002.
75. Western Development Commission. Biomass CHP market potential in the Eestern region: an assessment. Ballaghaderreen: Western Development Commission; 2008.
76. Department of Energy and Climate Change. A call for evidence on the role of gas in the electricity market. London: Department of Energy and Climate Change; 2012.
77. Innovations Orion. Walking the carbon tightrope: energy intensive industries in a carbon constrained world. London: The Trades Union Congress; 2014.
78. American Gas Foundation. Natural gas and energy price volatility: price volatility in today's energy markets. Oak Ridge: Oak Ridge National Laboratory; 2003.
79. Näsäkkälä E, Fleten SE. Gas fired power plants: investment timing, operating flexibility and abandonment; 2003.
80. Emery GW, Liu QW. An analysis of the relationship between electricity and natural-gas future prices. J Futures Mark. 2002;22(2):95–122.
81. Larsson R. A comparative case study of non technical barriers. Uppsala: Sveriges Lantbruksuniversitet; 2007.
82. Froning S, Piel E. Heat and power and district heating in the EU emissions trading scheme. Brussels: Euroheat & Power; 2007.
83. Finnish Energy Industries. Strategy for the district heating sector. Helsinki: Finnish energy industries, Helsinki; 2013.
84. E.ON UK. BERR heat call for evidence: response from E.ON UK. Coventry: E.ON UK; 2008.
85. Ecoheatcool. Reducing Europe's consumption of fossil fuels for heating and cooling. Brussels: Euroheat & Power; 2006.
86. Oxera. The cost of capital for heat distribution and supply: final report. The Hague: Energiekamer; 2009.
87. Local Energy Efficiency Consortium. Local energy efficiency project: project information booklet—phase one. London: Local Energy Efficiency Consortium; 2011.
88. Morris J. Transport biofuels in Sweden: Stakeholders' perceptions of the renewable energy. Lund: University of Lund; 2011.
89. Farkas A, Korhonen HP, Kuusela M. Benchmarking district heating in Hungary, Poland, Lithuania, Estonia and Finland: executive summary report. Budapest: Energy Regulators Regional Association; 2011.
90. European Investment Bank. The economic appraisal of investment projects at the EIB. Luxembourg: European Investment Bank; 2013.
91. Froning S, Piel E. Promotion of heating and cooling from renewable energies. Brussels: Euroheat & Power; 2006.

92. European Commission. Report from the Commission to the European Parliament, the Council, the European Central Bank, the European Economic and Social Committee, the Committee of Regions and the European Investment Bank. State of the single market integration 2013—Contribution to the annual growth survey 2013. Brussels: European Commission; 2013.
93. Bulmer S, Dolowitz D, Humphreys P, Padgett S. Policy transfer in European Union governance: Regulating the utilities. Oxon: Routledge; 2007.
94. Francoeur N. Critical assessment of Europe's energy market. 2008: a review of central and Eastern Europe, continental trends and EU legislation. Budapest: Energia klub; 2008.
95. Groznik A. Potentials and challenges in multi utility management. Abu Dhabi: European and Mediterranean Conference on Information Systems 2010; 2009.
96. Europe Cogen. Benchmarking report: status of CHP in EU member states. Brussels: Cogen Europe; 2006.
97. ESD. Sustainable energy communities and sustainable development: conclusions and recommendations. Wiltshire: ESD Ltd; 2005.
98. Cahn M. Liberalisation and its impact on municipalities in the participant countries and the UK: summary of report. Paris: International Energy Agency Demand-Side Management Programme; 2000.
99. Peters M, Fudge S, Wade J. Local authority perspectives on energy governance and delivery in the UK: a thematic assessment of enabling and restricting factors. Surrey: Centre for Environmental Strategy; 2013.
100. Commission European. Community guidelines on state aid for environmental protection (2008/C 82/01). Brussels: European Commission; 2008.
101. CrossBorder Bioenergy Working Group on District Heating. District heating: sector handbook. Jyväskylä: The Bioenergy Association of Finland; 2011.
102. The Sussex Energy Group. Centre on Innovation and Energy Demand. Brighton: University of Sussex; 2012.
103. Bale C, Bush R, Taylor P. Spatial mapping tools for district heating (DH): helping local authorities tackle fuel poverty. Leeds: Centre for Integrated Energy Research; 2014.
104. Sandén BA, Azar C. Near-term technology policies for long-term climate targets—economy wide versus technology specific approaches. Energy Policy. 2005;33(12):1557–76.
105. Persson U. Realise the potential! Cost effective and energy efficient district heating in European urban areas. Gothenburg: Chalmers University of Technology; 2011.
106. Possert T. Response to request for information "rate regulation". Graz: Energie Steiermark; 2013.
107. Umbach F. Critical energy infrastructure protection in the electricity and gas industries—coping with cyber threats to energy control centers. Berlín: Institut für Strategie Politik Sicherheits und Wirtschaftsberatung; 2010.
108. Li F. Spatially explicit techno-economic optimisation modelling of UK heating futures. London: UCL Energy Institute; 2013.
109. Entreprise Scottish. Scotland and the Central North Sea. London: Element Energy Limited; 2014.
110. Brinckerhoff Parsons. Decentralised energy masterplan for Westminster. London: Westminster City Council; 2013.
111. Toke D, Fragaki A. Do liberalised electricity markets help or hinder CHP and district heating? The case of the UK. Energy Policy. 2008;36(4):1448–56.
112. Suazo JA. Stages in project financing: a comparative analysis of independent power projects in three developing countries—India, Indonesia, and Peru. Massachusetts: Massachusetts Institute of Technology; 2001.
113. BRE, University of Edinburgh, Centre for Sustainable Energy. Research into barriers to deployment of district heating networks. London: Department of Energy and Climate Change; 2013.
114. Gartland D. Developing district heating in Ireland: why should it be developed and what needs to change?. Aalborg: University of Aalborg; 2014.

115. Cansino JM, Pablo-Romero M, Román R, Yñiguez R. Promoting renewable energy sources for heating and cooling in EU-27 countries. Energy Policy. 2011;39(6):3803–12.
116. Sletnes OH. EFTA surveillance authority decision of 9 May 2012 on the aid to Akershus Energi Varme AS for the construction and expansion of district heating and cooling infrastructure in Lillestrom, Strommen and Nitteberg. Brussels: EFTA surveillance authority; 2012.
117. Connor P. Documentation of consultation process on the proposed 2020/2030 RES-H targets in the United Kingdom. Exeter: University of Exeter; 2010.
118. Hughes L. Our winters of discontent: addressing the problem of rising home heating costs. Halifax: Canadian Centre for Policy Alternatives Nova Scotia; 2006.
119. IEA. CHP/DHC Country scorecard: Denmark. París: International Energy Agency; 2009.
120. Mautino L. Regulatory best practices applied to district heating. Riga: future of heat markets and DH pricing in Baltic countries and Poland; 2012.
121. Arecete JB. Aspectos contables de las colaboraciones público-privadas. Presupuesto y gasto público. 2006;45:199–214.
122. Brinckerhoff Parsons. Decentralised energy project borough-wide. London: London Borough of Islington; 2010.
123. Parsons Brinckerhoff Ltd. A district heating utility for the Tees Valley. London: Renew at the Centre for Processes Innovation; 2010. p. 10.
124. Tudway R. An enabling framework for district heating. London: Department of Energy & Climate Change; 2010.
125. Moss KJ. Heating and water design in buildings. 2nd ed. London: Taylor & Francis; 2003. p. 2003.
126. Høigaard M. NO—Meldal: District heating. London: Dods Parliamentary Communications Ltd.; 2012. http://www.publictenders.net/node/1692324. Accessed 24 Aug 2014.
127. OECD. Regulatory reform in Denmark: regulatory reform in the electricity sector. Paris: Organisation for Economic Co-operation and Development; 2000.
128. Commission European. Directive 2009/72/EC of the European Parliament and of the Council of 13 July 2009 concerning rules for the internal market in electricity and repealing Directive 2003/54/EC. Brussels: European Commission; 2009.
129. Dalla Rosa A. The development of a new district heating concept: network design and optimization for integrating energy conservation and reneweble energy use. Kongens Lyngby: Danmarks Tekniske Universitet; 2012.
130. Fudge S, Peters M, Wade J. Locating the agency and influence of local authorities in UK energy governance. Surrey: Centre for Environmental Strategy; 2012.
131. Poputoaia D, Bouzarovski S. Regulating district heating in Romania: legislative challenges and energy efficiency barriers. Energy Policy. 2010;38(7):3820–9.
132. Commission European. Energy markets in the European Union in 2011. Brussels: European Commission; 2012.
133. Pasoyan A. Regional urban heating policy assessment: Part I. Washington: Alliance to Save Energy; 2007.
134. ECRB. Treatment of the vulnerable customers in the Energy Community. Vienna: Energy Community Regulatory Board; 2013.
135. Westin P, Lagergren F. Re-regulating district heating in Sweden. Energy Policy. 2002;30 (7):583–96.
136. The Danish Energy Authority. Heat supply in Denmark: who, what, where and why. Copenhagen: The Danish Energy Authority; 2005.
137. Branson A. Extending district heating and CHP in central Leicester. Leicester: Leicester City Council; 2008.
138. Birnbaum A, Elleriis J. Best practice: district heating system (Copenhagen). New York: New York City Global Partners; 2011.
139. Nordic Energy Perspectives. The future of Nordic district heating: a first look at district heat pricing and regulation. Oslo: Nordic Energy Perspectives; 2009.

140. Control Energie. A better deal—whenever change brings opportunities. Vienna: Energie Control; 2010.
141. Ecoheat4eu. Support measures for DHC: Germany: Brussels: Ecoheat4eu; 2011. www.ecoheat4.eu/en/Country-by-country-db/Germany/Support-Measures-For-DHC. Accessed 26 Aug 2014.
142. Thollander P, Svensson IL, Trygg L. Analyzing variables for district heating collaborations between energy utilities and industries. Energy. 2010;35(9):3649–56.
143. Rhodes M. A new model for ECO. London: Encraft Ltd; 2014. www.encraft.co.uk/viewpoints/a-new-model-for-eco. Accessed 28 Aug 2014.
144. Radov D. Policy options to encourage energy efficiency in the SME and public sectors. London: Department for Environment, Food and Rural Affairs; 2006.
145. Brodies LLP. Making ESCOs work: guidance and advice on setting up & delivering an ESCO. London: London Energy Partnership; 2007.

Chapter 4
Reconciliation of Social Discount Rate and Private Finance Initiative: Application to District Heating Networks in the EU-28

The present would exercise dictatorship over the future by letting decisions be guided only by net present value.
—Anders Chr. Hansen, Roskilde University

The European Union needs, due to its energy dependency and high potential in reducing both primary energy consumption and carbon dioxide emissions, to promote the use of district heating networks and cogeneration as well as to remove the barriers that halt their spread, through the implementation of its community energy strategies. These technologies are among the few that, in the short to medium term, can contribute significantly to achieving in an economically feasible way the EU's energy and environmental goals.

Following the recommendations of the EU-28 for infrastructural investments in Western Europe, the economic feasibility of such schemes is assessed by conducting a cost-benefit analysis in which a social discount rate (SDR) of 3.5% is to be used. From a purely business perspective, once the net present values are calculated for different projects, the highest one will be chosen; however, from a government perspective, where many costs and benefits are difficult to quantify, the NPV can be only one of many criteria for decision-making.

These contradictory perspectives may impede the selection of projects involving the joint use of district heating networks and cogeneration. The purely business perspective would require "normal" discount rates for this type of project (usually between 10 and 12%), 10% being the lower limit for a project to be accepted. This research explains and proves why this apparent inconsistency is illusory, resulting from a faulty understanding of the role of governments in the promotion of the private sector, and how the 3.5% discount rate proposed by the EU can finally be used.

A. Colmenar-Santos et al., *District Heating and Cooling Networks in the European Union*, DOI 10.1007/978-3-319-57952-8_4

4.1 Introduction

As referenced in literature, the discounted cash flow analysis is regularly used in determining the feasibility of projects [1]. The advantage of this method is that, unlike the NPV and IRR methods, it compares well projects of various costs and useful lives [2]. To have a unique value representing the risk of the project [3], the discounted cash flow will finally give the NPV of the project [4]. As a rule, only if the NPV is positive for a given discount rate, it will be relevant to invest in a project [5]. If all the projects have negative net present values, as often it happens when the investments are infrastructural [6], then the least negative net present value will be selected [7].

The techniques of the discounted cash flow consider that a monetary unit today has not the same value as a currency at a future date [8]; this is because if the currency had been invested in a guaranteed investment, its value would have increased until, in exceptional cases, greatly exceed the typical return on government bonds for periods of 30 or more years [9]. As far as Western Europe is concerned, the growth of its economy in the long-term (30 years or more as a time horizon) is estimated to be about 2.5% per year [10, 11], which is why the profitability of government bonds for periods of 30 or more years is also 2.5% per annum [12].

The total present value of a cash flow time series would lead to the net present value (to its calculation, it would be necessary to discount the cash flow) [13–17] (4.1):

$$NPV = \sum\nolimits_{t=0}^{T} \frac{R_t}{(1+r)^t} \tag{4.1}$$

As a general rule, if the average investor wants his investment to be fully secured, he will not obtain more than that 2.5% annually. This is the reason why the recommendations for the Western European countries, proposed by the European Commission, indicate that those investments that are considered as infrastructural should not have a (social) discount rate higher than 3.5% (as is considered that such projects have a wide impact and are beneficial for society as a whole) [18]. This means that if a given government is convinced that an infrastructural investment can provide this discount rate of 3.5% and contribute to the achievement of some of their priorities (among which may be reducing pollutant emissions or the amount of imported fuels), the project should be undertaken.

It should be recognised that not all business projects are profitable, and some of them are only profitable for a certain period [19], which forces investors to diversify [20] and to finally get for their long-term investments a profitability similar to the one of the country as a whole, which is often similar to that of government bonds in the long term (30 years). As a general rule, private investors investing in low-risk projects, rather than government bonds, require an annual return of at least 6%, so assuming a typical 2% inflation and an additional premium risk of 1% [21] would result in a real yield of 3%.

Even though it is true that some European central banks have pushed policy rates below zero in the second half of 2014 and early 2015 (some bonds now offer negative yields) to combat deflation, this is a truly extraordinary, extremely rare phenomenon. Not even during the Great Depression were short-term rates negative [22]. This is a response to an exceptional financial situation that is not expected to continue long-term and, consequently, does not invalidate the real yield requirement of 3% for private investors.

The research proposed in this chapter is intended to correctly combine the binomial between social discount rates and private finance initiative, since from this reconciliation depends the compliance with Directive 2004/08/EC on promotion of cogeneration. Moreover, to our knowledge, scientific literature has not specifically addressed the issue of reconciliation between a social discount rate proposed by a supranational institution with the private finance initiative, so the approach and topic of this chapter are of paramount importance and truly innovative. The target of this research is clear: to propose measures allowing the PFI to embark on projects involving the joint use of district heating networks and cogeneration, accepting the discount rate proposed by the EU-28 Directorate General of Urban and Regional Policy for infrastructural projects. Note, however, that due to their nature, the measures proposed here will generally also apply to many low-carbon investments and even to many infrastructural investments.

In this chapter's first section, a first approach to the techniques used to assess the viability of projects is described. In addition, it mentions the special consideration that infrastructural projects should receive and discusses the need to reconcile the social discount rate proposed by the EU with the return required by private finance initiatives. In the second section, the theoretical background concerning the matters mentioned in the first section are analysed; subsequently, in the third section, measures for reconciling social discount rates and private financing initiatives are proposed. The fourth section is reserved for a discussion of the government's role in the selection of social discount rates; and finally, the fifth section will show the conclusions in which energy policy implications for the proposed measures will be presented.

4.2 Theoretical Background

By their very nature, the (social) discount rates are subjective parameters, one of their biggest problems being the fact that they are not observable or resulting from common sense [23–25].

The discount rate (despite having traditionally been a controversial issue) [26, 27] is an essential element in the economic evaluation of government policies and programs and projects [28]. In the '70s, and after the oil crisis that occurred in 1973, many countries saw the need to invest in energy research [26]. Previously (in particular, in 1967, with the publication of the report "The Economic and Financial Obligations of Nationalised Industries" [which continues to evolve]), the United Kingdom had already established a formal discount rate for projects that entail public-sector investment [29–33].

4.2.1 Cost Benefit Analysis

Although CBA analysis seems simple, this approach involves a number of practical complications [34–37]. First, the uncertainty regarding the potential proceeds from the assets means that obtained interest rates can be very different [37]—these uncertainties include risks related to the project and political and regulatory risks, as well as the uncertainty regarding the evolution of energy prices [38]. It should be noted that investors demand a high return (or risk premium) for assets associated with a high risk but accept much lower returns for assets involving safe investments [37–39].

4.2.2 Net Present Value Method

Even though this method did not emerge as a management tool to assess long-term investment projects until the '50s (formerly greater emphasis was placed on the recovery period) [40], was developed much earlier (in particular, since the late 16th century it was used for investments) [41], often being the critical factor for prioritising projects when resources are limited [31, 42]. If the NPV is positive (and no alternative is more attractive) [35], the project is efficient [23, 43–45] and therefore should be accepted [31, 46].

When it comes to private projects, market interest rates (which somehow estimate the reversal that can be sustained throughout the life of the project and determine whether an investment is economically feasible) [44] are the basis for calculating the discount rate [32, 46]. As it is already known, the high profits obtained in the financial markets are highly associated with increased risk of these investments and the fact that interest rates markets are based on activities with high productivity that often cause a high environmental pressure (in other words, high interest rates reflect an unsustainable system, therefore being unsustainable by themselves) [47].

4.2.3 Social Versus Private Discount Rates

Although there is no international agreement on its value, international practices recommend that the SDR vary between 2 and 7% for the EU and the US to evaluate infrastructural projects and policies converting future cash flows and benefits to their present equivalent [23, 46, 48–50]. It should be noted that it is remarkable that various governments have historically disagreed about the extent of the appropriate discount rate [51–53]. A low discount rate favours projects whose benefits occur in the long term [14, 35, 54], whereas high ones (its use implies that socially attractive projects are rejected and those whose benefits occur in the early stages are favoured) are detrimental in the long term [25]. When it comes to long-term projects, its variation greatly impacts the results of the projects [55, 56], leading to an intensification of the debate surrounding the numerical value of the discount rate [24].

In fact, existing literature, despite being extensive, is "inconclusive" [57], as it is possible to find authors who defend the use of a SDR equal to the interest of private investment and others who argue that this private opportunity cost does not put enough emphasis on the potential social advantages of infrastructural projects [27]. Certainly, the market interest rate does not adequately consider, among others, issues such as externalities or public goods, so some authors consider its adjustment imperative [58]. This debate is undoubtedly of importance because the discount rate has a great influence on investments that generate long-term costs and benefits [47, 57, 59–61].

It has been noted that although most economists consider it obvious that the social discount rate be positive, some "philosophical" authors claim that it "should" be zero [27, 59, 62] or even negative for certain projects with a strong social purpose [63] since in these cases, a positive discount rate could be a "polite expression for rapacity" [27]. In fact, many governments undertake infrastructural projects that, were a "conventional" discount rate be applied, would have a high probability of a negative NPV, justifying most public projects through externalities [25] and applying lower discount rates than conventional [26].

Some countries like the UK currently adopt a declining SDR for long-term projects to incorporate the effect of the uncertainty associated with the project and to increase the focus on future benefits [64]. However, this type of discount rate is not without problems. Some authors like Hepburn or Gollier argue that SDR will increase with the investment assessment day but will decrease "over time" [64, 65], by using the criterion of "logically equivalent expected net future value" instead of the "expected NPV" [65]. As a result, two apparently symmetrical and equally plausible ways of dealing with the uncertainty associated with prospective SDR yields completely contrary outcomes (this fact is known as the "Weitzman-Gollier puzzle") [61], which involves a temporarily inconsistent planning [64].

In recent decades, the use of the typical method for evaluating infrastructural projects (an annually constant discount rate) has been controversial (mainly in long-term projects) [66], with some agreement in the recent scientific literature about the idea that SDR varies [32, 67], which is necessary to achieve an intergenerational efficiency [24]. Although interesting, the fact is that these models (using declining discount rates) have not been widely used [68] due to, among other things, the fact that the mathematical specification of this equity is somewhat arbitrary [25]. Consequently, a constant discount rate remains the most widely used for evaluating public projects [69] and is recommended in most guidelines for economic analysis [70].

Certainly, a declining SDR can solve some problems only to create others [49]. The main flaw of any model with discount rates that diminish with the passage of time is its temporal inconsistency [20, 52, 67] (unlike varying SDR, constant discount rates are always temporally consistent) [28, 32, 49]. Such temporal inconsistency suggests that in many economic models, discounting with a constant discount rate is "the only dynamically consistent" discounting method [71].

For all of the above, this book saw it fit to take as valid the constant discount rate proposed by the EU-28 (3.5% for infrastructural projects located in Western Europe), proposing, in the next section ("Proposed Measures"), the measures that can reconcile the (constant) social discount rate with private finance initiatives.

4.3 Proposed Measures

Suppose that a government decides to go for a certain technology to achieve their energy and environmental goals. In such a case, and if the transaction costs of collective decisions are lower to the organisational and planning costs, then the projects associated with this technology should be conducted regardless of whether there is a need to manipulate the market interactions introducing some exogenous incentives.

In a typical public-sector project, the state assumes most of the risk associated [72]; meanwhile, in projects that are ruled under the terms of a well-structured private sector (such as those that occur in the current European economic orthodoxy), developers will be those who take the liabilities [72] and (undoubtedly) will require a discount rate higher than the 3.5% proposed by the EU-28 for infrastructural investments, making it necessary for the government to manipulate the market interactions and introduce the aforementioned exogenous incentives (in such a way that a higher income is ensured).

For example, projects that include the use of renewable energies receive in most parts of EU-28 different types of aid to make them attractive to investors [75]. Such aid generally comes from taxes or rates that the government requires utilities to impose on its customers [73], which ultimately has the same effect that is none other than promoting technologies that help to achieve the objectives of governments in a particular subject.

Therefore, if a particular government is convinced that one or more specific solutions are generally beneficial (and that they contribute to their overall strategy) [73], it shall manipulate market interactions and introduce the necessary exogenous incentives to obtain the discount rate required by the promoters and consequently manage that a particular project is carried out. It can provide between the typical 10 and 12% nominal [34, 74] or down till 6–10% for those projects with a remarkable social contribution [21, 74].

While quantitative easing has been occasionally used in Europe as a tool to reduce long-term interest rates, the fact is that its use dilutes the existing total wealth [75], being a not-at-all advisable option to carry out in projects involving the joint use of cogeneration and district heating networks. In this regard, and to encourage private finance initiative, a possible approach would be to allow by law that energy providers could get a (modest) internal rate of return.

It should be noted that the profitability of the project should be relatively attractive and not too variable—that is, there should be a close correlation between the costs and the resulting rate. This alternative does not really involve any subsidise but only a price regulation that would allow to obtain the typical (nominal) discount rate of other projects with similar associated risks that, as already mentioned, should be between 6 and 12% in order to be attractive (it should be noted that an improvement of the liquidity and a reduction in the overall risk of the country in economic terms will make investors accept lower discount rates).

When there are large upfront costs associated with the implementation of the necessary infrastructure for the generation and supply (which is the case, for example, with the networks of water and energy networks), in some circumstances, it is more efficient that a single company be the one that produces the good. For investors to embark on such projects and obtain an acceptable rate of discount, historically different European governments have ceded some of this powers to private companies—among which are the granting of monopolies de iure, powers over compulsory purchase of land and its use, and pre-granted licenses. In exchange for these powers, the government will require, among other requirements, that capital to invest is very high, which is a consequence of the "obligation to serve" that have this type of granted monopolies and that there is no abuse by companies because of a dominant position.

In general, monopolies are not desirable because, in the end, they are still a concentration of power that, without the use of certain mechanisms, can abuse of their privileged position [77]. But natural monopolies (which should be distinguished from simple monopolies) [76] are clearly preferable to the alternative that no company is willing to embark on certain projects and, as a result, cannot provide certain services. In this type of monopoly, fixed costs are very high compared with variable costs, and therefore the average costs are unaffordable for a small number of users, dropping those costs as the number of users increases. That is why a natural monopoly can provide a service at a lower cost than would be obtained if several companies were competing with each other.

Although it may be controversial, the fact is that if granted licenses are not limited via monopoly, there would be much less incentive for the necessary infrastructure to be built, and the conditions to obtain the needed low-risk environment for these type of projects would not be created. Note that although financing in the district heating sector is usually a problem in many European countries, this is not true in the case of Denmark where firms finance (completely) their investments in district heating networks and cogeneration plants at the lowest market rate, and banks compete to offer the best conditions as they perceive that security is high.

Table 4.1 Measures proposed to reconcile the social discount rate indicated by the EU for infrastructural projects with the private finance initiative

Identified barriers	Proposed measures
Social discount rate (3.5%) for infrastructural projects insufficient from the private sector perspective	The government should manipulate market interactions and introduce exogenous incentives so that the private sector will ensure increased income
Use of quantitative easing to reduce interest rates in the long term (which results in the dilution of wealth effect)	Allow by law that energy providers obtain a (modest) IRR. Profitability should be relatively attractive and not too variable (there should be a close correlation between the costs and the resulting fee)
	This measure does not imply any subsidies but only an adjustment of the prices that will allow (nominal) discount rates typical for other types of projects with similar risks associated
Large upfront costs associated with the implementation of the necessary infrastructure	In some circumstances, it is more efficient than a single company is the one that produces the good. In order for investors to embark on such projects and obtain an acceptable discount rate (and always taking into account the different situations of each member state), it is proposed that governments may transfer part of their powers to private companies (among those powers are awarding monopolies de iure, rights on compulsory purchase of land and its use, and pre-granted licenses)
	In return, in exchange for these powers, the government will require, among other requirements, that the capital investment is very high (which is a consequence of the "obligation to serve" that have these type of monopolies granted by the government) and that there is no abuse by companies because of a dominant position
Experience in other sectors with natural monopoly characteristics shows that utilities will not embark on entrepreneurial or innovative activities without an incentive to do so since they enjoy a privileged *status quo*	To reduce the risk associated with district heating networks, it is proposed creating anchor loads so that a load that would represent about 80% of the total capacity of the network can be guaranteed, being able to do this thanks to the signing of long-term contracts that significantly reduce the risk of oversizing potentially obsolete assets
There is a lack of legal certainty	Substantial raising of sanctions on those member states which apply retroactive measures (and, therefore, create legal uncertainty) on energy

Source Own elaboration

This perception of high security is due to, among other things, a stable national energy policy and the existence of clear responsibility functions and efficient decision-making in the companies concerned; so it is proposed as a measure to improve such security, raising substantially the amounts of penalties to those member states that apply retroactive measures (and, therefore, create legal uncertainty) on energy projects. In fact, in most cases and excepting the Nordic countries, investors are reluctant to invest in such projects primarily because the risks are too high and unquantifiable [78].

Table 4.1 summarises the proposals for reconciling the SDR proposed by the EU-28 for infrastructural projects (involving the joint use of district heating networks and cogeneration) with private finance initiatives.

4.4 Role of Government in the Selection of the SDR

The most important problem of the SDR theory lies in the fact that the justifications offered often vary depending on the analysis undertaken [42] and that it is possible to justify any reasonable discount rate through any "internally consistent" argument [25]. In fact, the truth is that the numerical value that is often assigned lacks an "explicit justification" and varies greatly depending on the school of economic thought, reflecting a failure to identify their theoretical base [35, 79].

This book has advocated that, when a government pays for its portfolio of projects (in particular those relating to infrastructural projects involving schemes where a joint use of district heating networks and cogeneration is used) through tax revenues or bonds, the SDR used to prove its attractiveness is the expected value of long-term Treasury bonds (typically 30 years); consequently being a social discount rate that varies within a range of 3–4% suitable for the countries of Western Europe. The European Union considers that this approach is appropriate for all types of infrastructural projects, and any other circumstances (such as adjustments of risk or opportunity costs) should not be included in the social discount rate used.

This chapter, far from trying to propose the "right" social discount rate, aims, assuming as appropriate the recommended discount rate indicated by the European Commission (3.5% for those infrastructural investments located in Western Europe), to explain why the conflict between the social discount rate and the IRR required by the private finance initiative is illusory (and derived from a mistake in understanding the role of government) and propose measures to allow this SDR to be finally used.

Really (and this is not a recent finding), calculating a "right" discount rate for any type of project is not possible [32, 80] and continues being debated [16, 23, 24, 28, 34, 38, 43, 47, 61, 69, 81–86] in the CBA literature [66] because there is no a purely economic justification for their choice [87]. This is especially true for long-term projects [56] for which the financial markets do not provide any guidelines [88].

The school of thought adopted by the EC (and therefore the accepted in this research) is the one that represents Samuelson and Vickrey (among others), which proposes that states have a particular capacity for reducing risks, so the risk of public projects will be lower than private ones [38, 40, 89].

Taking into account results from Chap. 5, and as an advance, it is possible to see that 30-year lifetime equipment and infrastructure (centralised gas boilers, investment in new wet systems, individual valve pressures, individual heat exchangers) constitute roughly 70% of the whole cost of the schemes whereas 60-year (transmission pipelines, heat accumulators, district heating networks) and 15-year (individual gas heaters, individual electric heaters) lifetime equipment and infrastructure constitute 15% of the annual investment costs each.

It is possible to see that discount rates have a different impact depending on the technology. For example, were a modification of the discount rate (from 3.5 to 7.5%) to be undertaken, the relative importance of 15-year and 30-year lifetime infrastructure would be reduced, respectively, by (roughly) 2 and 1.5%; for its part, were a modification of the discount rate from 3.5 to 7.5% for 60-year lifetime infrastructures undertaken, the relative importance of these infrastructures and technologies would increase by roughly 4%.

As expected, the net present value of these schemes diminished as the social discount rate increased. Finally, it was possible to see that, for a 60-year lifetime horizon for a project, the social discount rate proposed by the EU Directorate-General for Regional and Urban Policy at a constant 3.5% is virtually equivalent to the declining discount rates proposed by the governments of the UK and France (see Fig. 4.1).

Conventional thermal power stations evaluated in this book, the characteristics of the schemes assessed, and costs associated with their implementations are shown respectively in Tables 5.4, 5.5, and 5.6. Please bear in mind that the comments presented in this book are referred to the best options from an economic point of view.

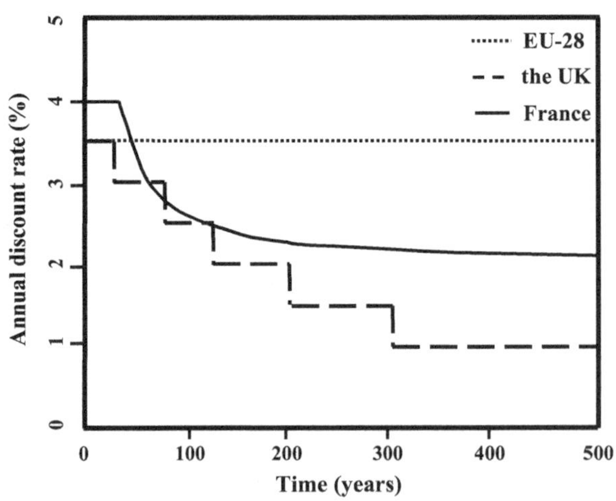

Fig. 4.1 Discount rates proposed by the EU-Directorate-General for Regional and Urban Policy, France and the UK [24, 32]

For the three cases conducted in Chap. 5 (a coal-based power plant located in Wilhelmshaven and two combined-cycle power plants located in Bristol and Cartagena), it is possible to see that the (theoretical) maximum discount rate which makes an investment feasible is, respectively, 16.6, 8.3, and 14.1% for the cases of Wilhelmshaven-Oldenburg, Bristol, and Cartagena. This great difference is due to the fact that the power plant located in Wilhelmshaven is a coal-based power plant with a lower electrical efficiency in condensing mode in comparison to combined-cycle power plants with a higher "cogeneration potential" and, consequently, lower annualised costs.

In the case of the power plant located in Cartagena, this very high "acceptable" discount rate is explained by the fact that if absorption units were used in the district heating network for cooling, annualised costs were to be greatly reduced (electricity prices are relatively high in Spain, constituting cooling expenses a considerable percentage of the electric energy consumption in summer). However, these two cases, with very high discount rates (for infrastructural projects) due to specific advantageous conditions, cannot be used in a generalised way as most infrastructural projects will be closer to the conditions of the district heating network proposed for Bristol (with an upper limit discount rate of 8.3% to make the project feasible).

Furthermore, it should be taken into account that this upper limit is only a "theoretical" one, as investors will not embark on a project that is expected to (no more than) equal the net present value of the current situation. The "real" lifetime of a project also plays a key role in its feasibility. For example, and, as always, restricted to the study conducted in Chap. 5, which supposed a project lifetime of 60 years for projects that entailed a joint use of district heating networks and cogeneration, the base cases were—even assuming a social discount rate of 3.5%—more economically advantageous than any other district heating solution if project lifetimes were to be reduced to between 34 and 42 years, depending on the district heating scheme evaluated. Uncertainty about the real lifetime of a project (in particular in infrastructural projects which are per se capital intensive) increases the risks associated with them and obliges governments to figure out what initiatives and measures (such as the ones proposed in Table 4.1) can be used to reconcile what happens in real life projects with the private finance initiative vision.

Consequently, each infrastructural project has its specific conditions and constraints and should be studied on an individual basis. Depending on the perceived risk of the country, an investor may be willing (or not) to invest in projects with similar net present values, and there is no possible to claim clearly, in which case, the concept of the social discount rate is no longer valid. Nevertheless, and as stated in the Introduction section of this chapter, a real yield of roughly 3% is demanded by investors for most "common" EU-28 economies. In case all (or most) of the proposed measures presented in Table 4.1 were implemented and in more advanced countries in terms of district heating networks such as Denmark, a social discount rate of 3.5% (the one proposed by the EU Directorate-General for Regional and Urban Policy) can be ultimately implemented as the government would have managed to create an environment of confidence, which makes possible the use of this SDR.

It should be taken into account that in the same way governments are proposed to be strongly sanctioned in case of legal uncertainty related to energy issues (see Table 4.1), investors are not supposed to be allowed to demand any internal rates of return different from the originally agreed ones; this why it is very important for a government to create the needed conditions (enforcing the measures proposed in Table 4.1) that create confidence in an investor.

Assuming the premise of some European countries such as France [90] or the new institutional economics [91], markets do not arise spontaneously and can be calibrated in the form of externalities [92]. It is precisely in this search for alternative measures where more has emphasised this chapter, proposing measures that are designed to succeed finally in attracting private finance initiative to infrastructural projects involving the joint use of district heating networks and cogeneration.

4.5 Conclusions

This chapter shows that there is no conflict between assessing infrastructural investments at a real discount rate of 3.5% and the fact that companies require a nominal disount rate of 6–12% for investing because governments can manipulate market interactions by introducing some exogenous incentives similar to those that currently exist for renewable energy in much of the EU-28 (through taxes or fees that governments force utilities to impose on their customers).

Electricity and gas license holders historically have had a competitive advantage over future competitors by virtue of their power regarding the compulsory purchase of land and its use and their ability to obtain pre-granted work licenses and be de iure monopolies. Such advantages are not normally enjoyed by operators of district heating networks and should be created and assigned to the promoters of projects that involve the joint use of district heating networks and cogeneration if it is desired that they be carried out.

For the expected lifetime of a district heating network/cogeneration project (typically evaluated for 60 years), there is virtually no difference between assessing this kind projects at the 3.5% (constant) discount rate proposed by the Directorate-General for Regional and Urban Policy and other declining discount rates for infrastructural projects such as those suggested by the governments of France and the UK for their respective countries. Unless it is recognised and remedied by the EU-28, the joint use of district heating networks and cogeneration will not have the intended outcome (as part of the program to combat climate change and reduce energy dependence) if all (or at least a combination of) the measures outlined in this book are not implemented to reconcile the social discount rate, as proposed by the European Commission, and private finance initiatives requirements.

References

1. Fanchi JR. Energy technology and directions for the future. 1st ed. Burlington: Elsevier; 2014.
2. Banerjee B. Cost accounting: theory and practice. 4th ed. New Delhi: Prentice-Hall of India; 2006.
3. Leach JC, Melicher RW. Entrepreneurial Finance. 5th ed. Stamford: Cengage Learning; 2014.
4. Energy Solutions. Interpreting clean power estimator results. Sacramento: California Energy Commission Renewable Energy Program; 2006.
5. Raghunathan V, Rajib P. Stock exchanges, investments and derivatives: straight answers to 250 nagging questions. 3rd ed. New Delhi: McGraw Hill Education India; 2007.
6. Damodaran A. Applied corporate finance. 3rd ed. Hoboken: Wiley; 2011.
7. Rogers M, Duffy A. Engineering project appraisal. 2th ed. Hoboken: Wiley-Blackwell; 2012.
8. Owen T. Assembly with robots. 1th ed. Berlin: Springer; 2012.
9. BAI. UK ground rents. Belgravia Alternative Investments Ltd, Hong Kong; 2014. http://www.belgravia-alternative.com/alternative-investments/uk-ground-rents/. Accessed 8 March 2015.
10. World Bank. Global economic prospects and the developing countries. 1st ed. Washington: World Bank; 2000.
11. Hannula H, Radosevic S, Von Tunzelmann N. Estonia, the New EU economy: building a Baltic miracle?. 1st ed. Aldershot: Ashgate Publishing; 2006.
12. BDE. Indicadores financieros: series diarias. Banco de España, Madrid; 2015. http://www.bde.es/webbde/es/estadis/infoest/si_1_2.pdf. Accessed 8 March 2015.
13. Van Wees JD, Bonte D, Genter A. Using decision support models to analyse the performance of EGS systems. Proceedings World Geothermal Congress 2010, Bali; 2012.
14. Yi ZF, Wong G, Cannon CH, Xu J. Can carbon-trading schemes help to protect China's most diverse forest ecosystems? A case study from Xishuangbanna, Yunnan. Land Use Policy. 2014;38:646–56.
15. Zhou Q, Mikkelsen PS, Halsnaes K, Arnbjerg-Nielsen K. Framework for economic pluvial flood risk assessment considering climate change effects and adaptation benefits. J Hydrol. 2012;414–415:539–49.
16. Claus K, Rousseau S. Public versus private incentives to invest in green roofs: a cost benefit analysis for Flanders. Urban For Urban Green. 2012;11:417–25.
17. Domenech L, Saurí D. A comparative appraisal of the use of rainwater harvesting in single and multi-family of the Metropolitan Area of Barcelona (Spain): social experience, drinking water savings and economic costs. J Clean Prod. 2011;19:598–608.
18. Directorate General Regional Policy. Guide to cost benefit analysis of investment projects. 1st ed. Brussels: European Commission; 2008.
19. Sander P. Ultimate guide to personal finance for entrepreneurs. 1st ed. Irvine: Entrepreneur Press; 2007.
20. Stevenson D. FT guide to exchange traded funds and index funds: how to use tracker funds in your investment portfolio. 2nd ed. Harlow: Pearson Education Limited; 2012.
21. CBFSAI. Financial stability report, 1st ed. Central Bank and Financial Services. Dublin: Authority of Ireland; 2005.
22. World Bank. Negative interest rates in Europe: a glance at their causes and implications. Washington: World Bank; 2015.
23. Levy N, Pauzner A. Government's credit-rating concerns and the evaluation of public projects. Journal of Public Economics. 2014;115:117–30.
24. Hepburn C, Koundouri P, Panopopoulou E, Pantelidis T. Social discounting under uncertainty: a cross-country comparison. J Environ Econ Manag. 2009;57:140–50.
25. Chen J. The nature of discounting. Struct Change Econ Dyn. 2012;23:313–24.
26. Almansa-Sáez C, Calatrava-Requena J. Reconciling sustainability and discounting in cost-benefit analysis: a methodological proposal. Ecol Econ. 2007;60:712–25.

27. Jetter K. Economic framework for decision making in biological control. Biol Control. 2005;35:348–57.
28. Hansen AC. Do declining discount rates lead to time inconsistent economic advice? Ecol Econ. 2006;60:138–44.
29. Kula E, Evans D. Dual discounting in cost-benefit analysis for environmental impacts. Environ Impact Assess Rev. 2011;31:180–6.
30. Weikard HP, Zhu X. Discounting and environmental quality: when should dual rates be used? Econ Model. 2005;22:868–78.
31. Thomopoulos N, Grant-Muller S, Tight MR. Incorporating equity considerations in transport infrastructure evaluation: current practice and a proposed methodology. Eval Program Plan. 2009;32(4):351–9.
32. Cruz-Rambaud S, Muñoz-Torrecillas M. Some considerations on the social discount rate. Environ Sci Policy. 2005;8:343–55.
33. Almansa C, Martínez-Paz JM. What weight should be assigned to future environmental impacts? A probabilistic cost benefit analysis using recent advances on discounting. Sci Total Environ. 2011;409:1305–14.
34. Howard G. Discounting for personal and social payments: patience for others, impatience for ourselves. J Environ Econ Manag. 2013;66:583–97.
35. Hauer E. Computing what the public wants: some issues in road safety cost-benefits analysis. Accid Anal Prev. 2011;43:151–64.
36. Parks S, Gowdy J. What have economists learned about valuing nature? A review essay. Ecosyst Serv. 2013;3:e1–10.
37. Howarth RB. Discount rates and sustainable development. Ecol Model. 1996;92:263–70.
38. Kesicki F, Strachan N. Marginal abatement cost (MAC) curves: confronting theory and practice. Environ Sci Policy. 2011;14:1195–204.
39. Sijtsma FJ, Van der Heide CM, Van Hinsberg A. Beyond monetary measurement: how to evaluate projects and policies using the ecosystem services framework. Environ Sci Policy. 2013;32:14–25.
40. Shaoul J. A critical financial analysis of the Private Finance Initiative: selecting a financing method or allocating economic wealth? Crit Perspect Account. 2005;16:441–71.
41. McHenry MP. Integrating climate change mitigation and adaptation: refining theory for a mathematical framework to quantify private and public cost-effectiveness, and C emissions for energy and development projects. Renew Energy. 2011;36:1166–76.
42. Frederick S. Valuing future life and future lives: a framework for understanding discounting. J Econ Psychol. 2006;27:667–80.
43. Esterhuizen JMC, Liebenberg GF. The use of indicators within a comprehensive impact assessment approach in three South African research programmes. Agr Ecosyst Environ. 2001;87:233–4.
44. Brown BB, Yiridoe EK, Gordon R. Impact of single versus multiple policy options on the economic feasibility of biogas energy production: swine and dairy operations in Nova Scotia. Energy Policy. 2007;35:4597–610.
45. Ledoux L, Turner RK. Valuing ocean and coastal resources: a review of practical examples and issues for further action. Ocean Coast Manag. 2002;45:583–616.
46. Fagiani R, Barquín J, Hakvoort R. Risk-based assessment of the cost-efficiency and the effectivity of renewable energy support schemes: certificate markets versus feed-in tariffs. Energy Policy. 2013;55:648–61.
47. Van den Bergh JCJM, Botzen WJW. Monetary valuation of the social cost of CO_2 emissions: a critical survey. Ecol Econ. 2015;114:33–46.
48. Kesicki F. What are the key drivers of MAC curves? A partial-equilibrium modeling approach for the UK. Energy Policy. 2013;58:142–51.
49. Guo J, Hepburn CJ, Tol RSJ, Anthoff D. Discounting and the social cost of carbon: a closer look at uncertainty. Environ Sci Policy. 2006;9:205–16.

50. Schucht S, Colette A, Rao S, Holland M, Schöpp W, Kolp P, et al. Moving towards ambitious climate policies: monetised health benefits from improved air quality could offset mitigation costs in Europe. Environ Sci Policy. 2015;50:252–69.
51. Brown-Gaddis E, Miles B, Morse S, Lewis D. Full-cost accounting of coastal disasters in the United States: implications for planning and preparedness. Ecol Econ. 2007;63:307–18.
52. Baum SD. Description, prescription and the choice of discount rates. Ecol Econ. 2009;69:197–205.
53. Fujii T, Karp L. Numerical analysis of non-constant pure rate of time preference: a model of climate policy. J Environ Econ Manag. 2008;56:83–101.
54. Abildtrup J, Audsley E, Fekete-Farkas M, Giupponi C, Gylling M, Rosato P, et al. Socio-economic scenario development for the assessment of climate change impacts on agricultural land use: a pairwise comparison approach. Environ Sci Policy. 2006;9:101–15.
55. Hansson SO, Lilieqvist K, Björnberg KE, Johansson MV. Time horizons and discount rates in Swedish environmental policy: who decides and on what grounds? Futures. 2015;76:55–66.
56. Johansson-Stenman O, Sterner T. Discounting and relative consumption. J Environ Econ Manag. 2015;71:19–33.
57. Rabl A. Discounting of long-term costs: what would future generations prefer us to do? Ecol Econ. 1996;17:137–45.
58. Padilla E. Intergenerational equity and sustainability. Ecol Econ. 2002;41:69–83.
59. Harford JD. Stock pollution, child-bearing externalities, and the social discount rate. J Environ Econ Manag. 1997;33:94–105.
60. Wise RM, Van Wilgen BW, Le Maitre DC. Costs, benefits and management options for an invasive alien tree species: the case of mesquite in the northern Cape, South Africa. J Arid Environ. 2012;84:80–90.
61. Gollier C, Weitzman M. How should the distant future be discounted when discount rates are uncertain? Econ Lett. 2010;107:350–3.
62. Birol E, Koundouri P, Kountoutis Y. Assessing the economic viability of alternative water resources in water-scarce regions: combining economic valuation, cost-benefit analysis and discounting. Ecol Econ. 2010;69:839–47.
63. Al-Amin AQ, Rasiah R, Chenayah S. Prioritizing climate change mitigation: an assessment using Malaysia to reduce carbon emissions in future. Environ Sci Policy. 2015;50:24–33.
64. Traeger CP. Discounting under uncertainty: disentangling the Weitzman and the Gollier effect. J Environ Econ Manag. 2013;66:573–82.
65. Hepburn C, Groom B. Gamma discounting and expected net future value. J Environ Econ Manag. 2007;35:99–109.
66. Defrancesco E, Gatto P, Rosato P. A "component-based" approach to discounting for natural resource damage assessment. Ecol Econ. 2014;99:1–9.
67. Karp L. Global warming and hyperbolic discounting. J Public Econ. 2005;89:261–82.
68. Cairns J. Development in discounting: with special reference to future health events. Resour Energy Econ. 2006;28:282–97.
69. Luisetti T, Turner RK, Bateman IJ, Morse-Jones S, Adams C, Fonseca L. Coastal and marine ecosystem services valuation for policy and management: managed realignment case studies in England. Ocean Coast Manag. 2011;54:212–24.
70. Severens JL, Milne RJ. Discounting health outcomes in economic evaluation: the ongoing debate. Value Health. 2004;7(4):397–401.
71. Settle C, Shogren JF. Hyperbolic discounting and time inconsistency in a native-exotic species conflict. Resour Energy Econ. 2004;26:255–74.
72. Vedavalli R. Energy for development: twenty-first century challenges of reform and liberalization in developing countries. 1st ed. London: Anthem Press; 2007.
73. Kirkham R, Martin J. The creation of an English Public Services Ombudsman: mapping a way forward. London: Democratic Audit; 2014.
74. University of Southern California. Economic and financial assessment. Plan Mark. 2001;4 (1):15–23.

75. Knott P. Ouch!: ignorance is bliss, except when it hurts—What you don't know about money and why it matters. 1st ed. Harlow: Pearson; 2012.
76. Cohen B. The Future of Money. 1st ed. Princeton: Princeton University Press; 2006.
77. Hillman AL. Public finance and public policy: responsibilities and limitations of Government. 2nd ed. Cambridge: Cambridge University Press; 2009.
78. Canada PPP. Energy-from-waste: sector study. Ottawa: PPP Canada; 2014.
79. Parkinson A, Guthrie P. Evaluating the energy performance of buildings within a value at risk framework with demonstration on UK offices. Appl Energy. 2014;133:40–55.
80. Al-Kasim F, Soreide T, Williams A. Corruption and reduced oil production: An additional resource curse factor? Energy Policy. 2013;54:137–47.
81. Scarborough H, Sahin O, Porter M, Stewart R. Long-term water supply in an Australian coastal city: Dams or desalination? Desalination. 2015;358:61–8.
82. Gollier C. Expected net present value, expected net future value, and the Ramsey rule. J Environ Econ Manag. 2010;59:142–8.
83. Elliston B, MacGill I, Diesendorf M. Least cost 100% renewable electricity scenarios in the Australian National Electricity Market. Energy Policy. 2013;59:270–82.
84. Jaeger WK. Is sustainability optimal? Examining the differences between economists and environmentalists. Ecol Econ. 1995;15:43–57.
85. Makridis C. Offshore wind power resource availability and prospects: a global approach. Environ Sci Policy. 2013;33:28–40.
86. Speck S. A neo-Austrian five process model with resource extraction and pollution abatement. Ecol Econ. 1997;21:91–103.
87. Chapman CB, Ward SC, Klein JH. An optimized multiple test framework for project selection in the public sector, with a nuclear waste disposal case-based example. Proj Manag. 2006;24:373–84.
88. Jouini E, Napp C. How to aggregate experts' discount rates: an equilibrium approach. Econ Model. 2014;36:235–43.
89. Tirelli M. The evaluation of public investments under uncertainty. Res Econ. 2006;60:188–98.
90. Andersen SS, Eliassen KA. Making Policy in Europe: the Europeification of National Policy-making. 1th ed. Thousand Oaks: SAGE Publications Ltd.; 1993.
91. Rosser A. The politics of economic liberalization in Indonesia: state, market and power. 1st ed. London: RoutledgeCurzon; 2002.
92. Hodge ID, Adams WM. Neoliberalisation, rural land trusts and institutional blending. Geoforum. 2012;43:472–82.

Chapter 5
Evaluation of the Cost of Using Power Plant Reject Heat in Low-Temperature District Heating and Cooling Networks

Prejudice is a burden that confuses the past, threatens the future and renders the present inaccessible.

—Maya Angelou

The purpose of this chapter is to assess the economic impact following as a result of the conversion of conventional stations to cogeneration plants connected to a system for distributing heat and refrigeration. The analysis will be carried out through a financial evaluation in order to assess yearly variations of the whole scheme expenditure.

In particular, yearly capital and operating costs of supplying cold and heat to urban areas that represent weather conditions in northern, central, and southern Europe (Oldenburg-Wilhelmshaven, Bristol, and Cartagena, respectively) were studied. In the base case scenario, today conditions are represented; they will be compared with alternative scenarios in which existing conventional thermal power stations will be converted into cogeneration plants.

5.1 Introduction

As a general rule, the newly-built thermal generating stations are designed as condensing power stations or as cogeneration plants [1]—most of them having the flexibility to switch back to condensing mode on a seasonal basis without the loss of efficiency [2] for the latter [3]. The technical difference between an electricity-only power plant and a cogeneration plant suitable for district heating networks lies in the fact that with the latter, there should be some way to condense the mixture of gas and saturated vapour at a superior temperature [4]. It is accomplished in cogeneration plants with extraction-condensing turbines that use the exhaust steam from the low-pressure turbine stage in a heat exchanger [5]. In this type of cogeneration plant, there is a possibility of modifying the amount of steam passed to the condenser (see Fig. 5.1). The more steam enters the condenser, the smaller the amount of electricity generated (due to the fact that a lower amount of gas and

A. Colmenar-Santos et al., *District Heating and Cooling Networks in the European Union*, DOI 10.1007/978-3-319-57952-8_5

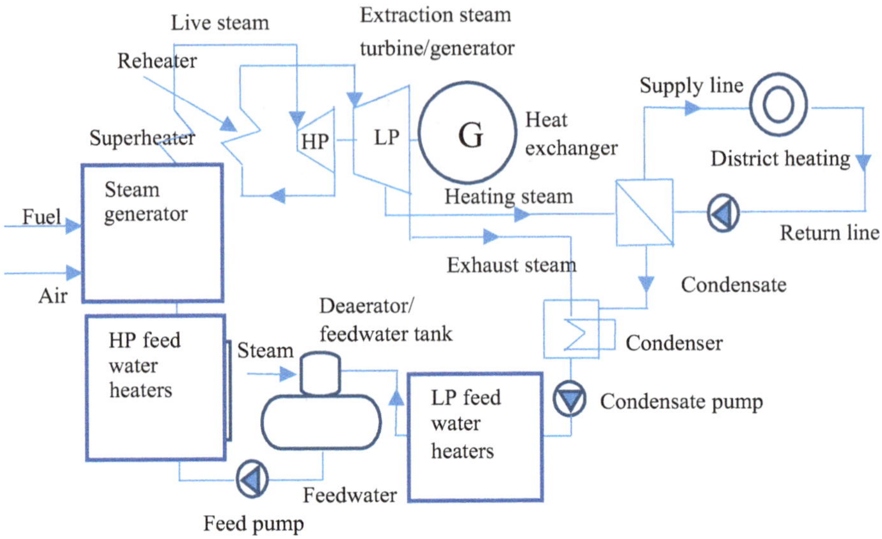

Fig. 5.1 Schematic diagram of a cogeneration plant for electricity generation and district heating based on an extraction-condensing turbine [5]

saturated vapour is leaving the low-pressure turbine). Extra information about extraction-condensing and backpressure turbines is provided in Appendix I.

Depending on the type of plant, power rating, and location, the extra cost for the cogeneration version varies between 10 and 45% [6]. If the aim is having a cogeneration plant as efficient as possible, the replacement of the low-pressure turbine [7] (or even replacement of the high-pressure one) [8] will likely be compulsory [9]; it should be noted that although this is an expensive solution, it will always be much cheaper than building a new power station [10].

In those countries where the technology of cogeneration and district heating networks is more advanced (such as in the case of Denmark), many of the units of thermal power plants converted into cogeneration plants have been (or will be) replaced by more modern units located on the same site, dismantling the old units [11] or using them as a reserve. In any case, it should be noted that other non-technical factors, such as the energy planning regime, will also come into play and that under no circumstances can all existing power plants be converted into cogeneration plants, having the actions to be taken considered individually.

It should be noted that, assuming that the total conversion (of all units) of a thermal power plant provides a higher heat than needed by the loads, it is possible to proceed to a partial conversion of the plant [12] at an obviously inferior cost. For example, considering the combined cycle power plant of El Fangal in Cartagena, Spain [13], it would be possible to convert only one of the three units of 400 MW. If the destination of the heat were low-temperature district heating networks (with a lower energy demand than conventional district heating networks), the number of units to convert would obviously decrease [14].

The conversion is achieved by a modification/replacement of both the LP turbine [15] and the condenser [16]. Alternatively, one unit may go into a partial modification to provide a warmer heat, extracting from the space between the high pressure turbine and the low pressure turbine [17] heat at a high temperature and pressure [18]. This solution would have a lower overall efficiency than the replacement of the low-pressure turbine [19]. The best option in any case depends on a particularised assessment of the generating station and on the end-use purpose and heat requirements [14].

In Fig. 3.3, the current spread of district heating networks in the EU can be seen. Please note that with the exception of the Nordic countries and some exceptional cases, most of these systems cover a very limited part of the city they cover. To understand the importance of using low-temperature district heating networks, the Carnot cycle has to be considered. According to this cycle, the greater the difference between the inlet temperature and the heat rejection temperature (the one that leaves the low-pressure turbine), the greater the efficiency of the power plant [19]. This is the reason why power plants tend to be located as close as possible to the sea (or to a river) and why it is important, in order to obtain the highest coefficient of performance, for the supply and return temperatures of the district heating networks to be as low as possible [20].

In the scientific literature, it is possible to find a large number of research papers exploring the economic and/or environmental impacts of the joint use of cogeneration and district heating networks. Among the most notable may be those carried out by Lončar and Ridjan, who focused on the environmental, energy efficiency, and economic issues of the joint introduction of district heating networks and cogeneration for a Croatian town [21]; those conducted by Margaritis et al., who examined two district heating networks of Greece based on fossil fuel (lignite) regarding alternative options for covering a part of the nearby cities' thermal loads (Kozani and Ptolemaida) [22]; the research undertaken by Trygg and Amiri who analysed, from a European system perspective, the most cost-effective technology for the production of cooling [23]; or the investigation conducted by Persson and Werner, who carried out an aggregated quantitative analysis of current average EU heat recovery levels in district heating systems [24].

However, the potential repercussions that weather conditions have on the feasibility of the joint use of district heating networks and cogeneration have not been given the same attention, so a study that addresses them is necessary. From a deeper survey of grey literature [25–33] and updated literature [34–60] related to the topic addressed here, it was possible to find that—even though there are plenty deal of different approaches—this book undoubtedly contributes to the pool of existing knowledge by giving a weather centered system perspective (to our knowledge, so far not addressed for district heating networks using the reject heat from thermal power plants). By performing this thorough literature review, we ensure the originality of the idea and method here presented. From this line of analysis, a much clearer insight of weather repercussions on the feasibility of similar-sized district heating networks is gained (so far not explicitly shown to our knowledge in any scientific paper).

Furthermore, the proposed method here presented improves analysis in the sense that a systematic and easy approach (neither strong mathematical skills nor specific software are required) manages to make comparisons among different district heating networks systems regardless of their sizes and locations.

This first section briefly discussed the possibilities of conversion of conventional thermal power plants into cogeneration plants and identified the EU-28 cities that have district heating networks (regardless of system size); in the second section, the method used to conduct the research will be exposed; in the third section, case studies on the implementation of low-temperature district heating networks in three cities of the EU-28 with different weather conditions will be presented; subsequently, in the fourth section results for case studies of selected cities will be shown. The fifth section is reserved for discussion and conclusions, where the economic and environmental consequences resulting from the implementation of the outlined projects in the third and fourth sections are presented. Finally, extra information about extraction-condensing and backpressure turbines (Appendix I) and heat accumulator storage (Appendix J), and supporting data (Appendix K) for the tables presented in the Results section of this chapter are provided in the Appendices section.

5.2 Materials and Methods

Knowing the needed energy to provide heating and cooling to a specific number of consumers and from various techno-economic parameters of the selected conventional thermal power plants (which will also be evaluated assuming them were converted to cogeneration plants), the three base cases will be sustained; they will include costs and revenues of the plants as well as the expenses associated with the loads of consumers utilising equipment available as of today. This assessment will include the cost of capital, renovation costs, operational and maintenance costs, and revenues from electricity generation. Please bear in mind that, for all the chapter, the heat we are interested in and refer to with the term “reject heat” is the waste heat cogenerated from the “electrical generators” of the thermal power plants and not the un-recovered heat of a previously existing cogeneration plant.

For the evaluation of the cases in which the conventional thermal power station is converted into a cogeneration plant, the cost of this conversion, infrastructure investment (district heating/cooling networks), buildings adaptation costs, and electricity prices are taken into account (in case heat accumulators that can cope with a demand for one to three days are included, it is possible to presume a power optimum instant selling) [61].

To make the economic analysis, annual costs of the systems for the specified cases, net electricity (difference between the electricity measured at the busbars of the powerplant and its consumption), primary energy used, and carbon dioxide emissions (calculated in each case directly from the primary energy) were compared.

5.2.1 Annual Costs

Annual costs of the system were calculated for each case using (5.1) [62–65].

$$C = AC_t + m \cdot X + v \cdot X + f - REV \tag{5.1}$$

5.2.1.1 Annualised Capital Costs

The annual outgoing aggregate of all future cash flows of the facilities are assessed employing Eqs. (5.2) and (5.3) [63, 66, 67]:

$$AC_t = \sum\nolimits_{t=1}^{T} A\,(n,d) \cdot c_t \cdot X \tag{5.2}$$

$$A\,(n,d) = \frac{d \cdot (1+d)^n}{(1+d)^n - 1} \tag{5.3}$$

Because district heating is an infrastructural investment [68], and in order to be consistent with the Directorate-General for Regional and Urban Policy proposal, a (social) discount rate of 3.5% was chosen because it was considered that such projects have a positive impact on the collective as a whole [69]. The result of using other higher discount rates (5.5 and 7.5%) will be checked later in a sensitivity analysis. The reason for choosing these discount rates is that they are, on the one hand, the social discount rates proposed by the EU-28 for competitive regions (3.5%) and for cohesion regions (5.5%) [70] and on the other hand, 7.5% is one of the most widely used discount rates for the evaluation of such projects [71–73].

In this book, it will be assumed that facilities are renovated, which obviously will have a positive impact in their lifetime or technical life [74] and will involve higher annual maintenance costs due to this economic life extension [75]. In addition to this, it is assumed that the inflation and the discount rate will remain constant over the useful life of the project, so the levelised cost of energy can be expressed in €/MWh, thereby eliminating the impact of inflation [76] and enabling the use of a constant currency (constant euros) tied to the current year. Ordinarily, life-cycle costing techniques use (as in this book) a real discount rate, which is entirely consistent with the constant currency analysis carried out in this book. It should be noted (previous studies at European level prove so) that the feasibility of the proposed actions in this book does not show a "high sensitivity" to changes in oil prices, natural gas, and coal (being "particularly insensitive" to the latter) [77], so the sensitivity analysis is performed only for the discount rate to be used.

5.2.1.2 Fixed Running and Preservation Expenses

Generating stations are regularly renovated [78] to extend their useful lives [79, 80], so it has been considered appropriate to add to the annual fixed costs of operation and maintenance (which average around 2% of the investment cost) [81, 82] an amount equal to them to satisfy renovation costs. It should be noted that the estimation of the latter presents a significant challenge [83] and that, depending on the type of plant, maintenance performed, age, and the country in which the plant is located, annual fixed costs of operation and maintenance may be anywhere from negligible [84] to representing 15% [85] or more [86] of the investment cost of the plant (due to these circumstances, it was considered that 2% is a quantity general enough to represent the renovation costs that have to be faced by the typical power plants of Western Europe).

5.2.1.3 Variable Running and Preservation Expenses

Apart from the fixed operating costs, it has been considered a variable running and preservation expense in a balanced relation to produced electricity [87].

5.2.1.4 Combustible Expenses

Mainly using data obtained from Europe's Energy Portal, different fuel prices for the residential sector (utilising the prices offered to households of each EU country) and the industrial sector (assuming here that the power plant will acquire the combustible as an industrial user) will be adopted.

5.2.1.5 Revenues Resulting from the Sale of Power

It has been accepted that sales of electricity take place in a cash market, taking into account data obtained from the electricity operators—EEX (for Oldenburg-Wilhelmshaven), Elexon (for Bristol), and OMEL (for Cartagena). Supposed prices can be seen in Table 5.1.

It should be noted that, in the analysis carried out in this chapter, a cost boundary around the system (power station, heat consumer, power consumer) is drawn and then added up the costs of fuel and capital under the combined heat and power district heating schemes versus the current (conventional) scheme. This means there is no need to allocate a specific cost of heat since the aim of the analysis is looking at the total cost flows into or out of the system—this view is supported by the senior energy experts consulted (see the Acknowledgements section).

Table 5.1 Electricity prices assumed in spot markets and to domestic consumers, and coal and natural gas prices (for both industrial and domestic consumers) to Oldenburg-Wilhelmshaven, Bristol, and Cartagena [88–92]

City	Concept	Price	Units
Oldenburg-Wilhelmshaven	Assumed carbon price[a]	16	€/MWh
Bristol	Assumed price of natural gas for industrial consumers	23	€/MWh
Cartagena	Assumed price of natural gas for industrial consumers	24	€/MWh
Oldenburg-Wilhelmshaven	Assumed price of electricity for the spot market operated by EEX	52	€/MWh
Bristol	Assumed price of electricity for the spot market operated by Elexon	57	€/MWh
Cartagena	Assumed price of electricity for the spot market operated by OMEL	51	€/MWh
Oldenburg-Wilhelmshaven	Assumed price of natural gas for domestic consumers	67	€/MWh
Bristol	Assumed price of natural gas for domestic consumers	60	€/MWh
Cartagena	Assumed price of natural gas for domestic consumers	85	€/MWh
Oldenburg-Wilhelmshaven	Assumed price of electricity for domestic consumers	298	€/MWh
Bristol	Assumed price of electricity for domestic consumers	191	€/MWh
Cartagena	Assumed price of electricity for domestic consumers	225	€/MWh

[a]Note that the thermal power station of Wilhelmshaven (E.ON) uses coal as fuel, so, unlike the thermal power plants of Seabank (Bristol) and El Fangal (Cartagena), which use natural gas, it is not necessary to include the price of natural gas for industrial consumers. Meanwhile, due the fact that domestic consumers from Oldenburg-Wilhelmshaven use natural gas (like Bristol and Cartagena), in this case, obviously, its assumed price has been shown

5.2.2 Conversion and Emission Factors

Aside from the coal and natural gas used by thermal power plants and domestic consumers, the primary energy demand includes the electricity sent to the grid [93].

The annual energy required to generate the supply of the energy carrier electricity is calculated according to the typical efficiencies of the most extended power generating technologies available today and on the consideration of a losses in transmission and distribution networks of 9% for the electricity mix of each country (it should be noted that the three cities studied are located in Western Europe, so 9% is reasonable) [94–97]. The same, presented in Table 5.2, is based on statistics from *The Shift Project Data Portal* for 2012. The primary energy conversion factors calculated for the plants evaluated [98], the assumed efficiencies [99, 100] and the emission factors used (note the divergence of the same depending on the country studied and bibliography used) [98, 101–103], are presented in Table 5.3.

Table 5.2 Electricity mix for the year 2012 for the countries of the cities studied

Energy source	Oldenburg-Wilhelmshaven (Germany) (%)	Bristol (United Kingdom) (%)	Cartagena (Spain) (%)
Natural gas	11.0	28.0	25.0
Coal	47.0	40.0	19.0
Oil	2.0	1.0	5.0
Nuclear energy	16.0	19.0	21.0
Biomass	8.0	5.0	2.0
Hydropower	4.0	2.0	7.0
Wind power	8.0	5.0	18.0
Solar photovoltaics	4.0	0.0	3.0

Source The Shift Project Data Portal

5.3 Calculation

For each of the cities of Oldenburg-Wilhelmshaven, Bristol, and Cartagena, its heating and cooling demands have been taken into account (these demands obviously depend on their respective weather conditions). Each of the cases considered are represented as follows:

(a) *Current Situation* (*Base Case*)

In the current situation, electricity is sold in the cash market by the generating stations without providing heat to any system for distributing it. Refrigeration is provided through conventional vapour-compression refrigerators and evaporative coolers.

To assess this case, the following technologies have been considered:

- The conventional thermal power plant assigned to Oldenburg-Wilhelmshaven has been the coal power plant at Wilhelmshaven (E.ON); for Bristol, the assigned plant has been the combined cycle power plant of Seabank; meanwhile, the thermal combined cycle power plant of El Fangal has been chosen to conduct the evaluation for Cartagena.[1]
- Individual heating and cooling equipment using gas and electricity.

[1]To carry out the case studies and in the choice of the type of thermal power plant (which obviously will impact on the type of fuel used), the sample has tried to be as representative as possible. Although there are some exceptions (such as the power plant of Frimmersdorf, which is located 35 km from Cologne and that currently uses its residual heat in the district heating network of Grevenbroich, also known as the "Capital of Energy") [109], the fact is that lignite power plants tend to be located far from the thermal loads of cities to be normally located near lignite mines. A similar situation occurs with nuclear power plants, which tend to be located away from big cities, so they will not be selected for evaluation.

Table 5.3 Primary energy conversion factors, efficiency and emission factors used for the power plants studied

Power plant	Type	Fuel used	Electrical efficiency in condensing mode (%)	Energy used to generate electricity (TWh)	Output power (TWh)	Losses in transmission and distribution networks (TWh)	Conversion factor	Emission factor (tCO_2/MWh)
Wilhelmshaven (E.ON)	Coal power plant	Coal	46	9.51	4.37	0.39	2.39	0.50
Seabank (Bristol)	Combined cycle power plant	Natural gas	55	12.18	6.70	0.60	1.99	0.24
El Fangal (Cartagena)	Combined cycle power plant	Natural gas	55	12.81	7.04	0.63	1.99	0.24

Source Own elaboration

(b) *COGEN-a1 Case*

In this case, the conventional thermal power plants have been considered as converted into cogeneration plants through their conversion and not through the construction of completely new plants; the investment associated with the infrastructure necessary to enable the implementation of low-temperature district heating networks has also been taken into account. Unlike COGEN-b and COGEN-c, cooling is carried out through electrical refrigeration equipment.

Introducing a thermal energy storage, it is plausible to suppose that the cogeneration plant will provide the same number of hours of electricity that the conventional generating station of the current situation has. It should be noted it is a cautious supposition, as cogeneration plants typically have an electrical production duration higher than conventional power plants [104], which adds an additional benefit to the system. Please note that this additional heat modifies the economic conditions of operation of the plant due to the reduction of operational costs [105] and increased profitability [106].

The option COGEN-a1 has been compared to the Base Case alternative performing a sensitivity analysis. In the Base Case alternative, current heating appliances are kept in buildings, not producing any improvement in those nor therefore investment.

For the calculation of the case COGEN-a1, the following technologies and considerations have been taken into account [107]:

- Costs of retrofitting existing conventional power plants into cogeneration plants; transmission pipelines from power plants to district heating networks; district heating networks of cities; and accumulators in district heating networks.
- Centralised natural gas boiler systems for peak demand; connection costs of individual housing and heat exchangers; and individual cooling appliances.

(c) *COGEN-a2 Case*

The COGEN-a2 case is similar to COGEN-a1, but building a new cogeneration plant instead of converting the conventional thermal power station.

(d) *COGEN-b Case*

Unlike the case COGEN-a1, COGEN-b assumes (individual) absorption units use the district heating network for cooling.

For the calculation of the case COGEN-b, the following technologies and considerations are taken into account:

- Costs of retrofitting existing power plants into cogeneration plants; transmission pipelines from power plants to district heating networks; district heating networks of cities; and accumulators in district heating networks.
- Central heating (boilers) using natural gas for peak demand; cost of heat exchangers and connection to buildings; and absorption units used as individual cooling devices.

(e) *COGEN-c Case (trigeneration)*

Unlike the cases COGEN-a1-a2, in COGEN-c, it is assumed that the cooling is obtained through a district heating network and integrated absorption chillers (which refrigerate the district energy scheme utilising the waste heat cogenerated from the "electrical generators"). It is assumed that absorption units are located between the transmission pipelines and the district heating networks of the cities.

In the case COGEN-c, the following technologies and considerations are assumed:

- Cost of conversion of existing conventional power plants into cogeneration plants; transmission pipelines from power stations to district heating networks; district heating networks; district cooling networks; gas central heating (boilers) for peak demand; accumulators in district heating networks; and absorption refrigeration units for district cooling.
- Cost of heat exchangers and connection to buildings; individual cooling appliances.

5.3.1 Power Plants and District Heating and Cooling Networks

Please note that in this book, the only conventional thermal generating stations that have been evaluated are those stations that, apart from meeting the requirement of being able to easily fit into the climatic conditions of northern, central, and southern Europe, meet certain restrictions, such as being located at a manageable distance from the nearest town, and that can be easily retrofitted into cogeneration plants at a reasonable cost.

In Table 5.4, selected conventional thermal plants are exposed, showing the techno-economic parameters of those converted into cogeneration plants and those of new construction in Tables 5.5 and 5.6, respectively.

The most important characteristics of the equipment and infrastructure required for the district heating network (such as lifespan, availability, thermal and refrigeration performance, or running expenses) are included in Table 5.6. Much of the

Table 5.4 Conventional thermal power stations evaluated

Power plant	Type	Fuel	Rated power (MW)
Wilhelmshaven-E. ON [109]	Steam turbine	Coal	757 (1 × 757 MW)
Seabank (Bristol) [108]	Gas turbine combined cycle	Natural gas	1140 (1 × 755 MW + 1 × 385 MW)
El Fangal (Cartagena) [13]	Gas turbine combined cycle	Natural gas	1200 (3 × 400 MW)

Source Adapted from [13, 108, 109]

Table 5.5 Technical and financial data from the evaluated generating stations

	Rated power (MW)	Thermal power (MWt)	LS (years)	AVAIL	η_e	η_{BP}^{a}	η_h	Heat to electricity ratio	CEH[b]	c_t^c (€/kW)	m·X[d] (€/kW)
Existing plant—Oldenburg-Wilhelmshaven	757		60	0.95	0.46						72
Existing plant—Bristol	1140		60	0.94	0.55						31
Existing plant—Cartagena	1200		60	0.94	0.55						31
Cogeneration conversion—Oldenburg-Wilhelmshaven	757	931	60	0.95	0.46	0.39	0.48	1.23	0.11	400	72
Cogeneration conversion—Bristol	1140	878	60	0.94	0.55	0.52	0.40	0.77	0.11	200	31
Cogeneration conversion—Cartagena	1200	923	60	0.94	0.55	0.52	0.40	0.77	0.11	200	31
New cogeneration plant—coal			60	0.95	0.46	0.39	0.48	1.23	0.11	1800	72
New cogeneration plant—gas			60	0.94	0.55	0.52	0.40	0.77	0.11	775	31

Source Adapted from [84, 99, 110–120]

[a]It was considered an overall efficiency of 90% in backpressure mode [126]

[b]Ratio between the electricity gained and the heat lost [110, 111] (it was assumed that thermal energy recovery is produced at 80 °C) [112, 113]. In the case of more widely used temperatures, such as those close to 130 °C [114], this ratio would rise to about 0.2 [115]

[c]See [116–118]

[d]It is considered that the costs of operation and maintenance are 2% per year of the investment costs [119], adding another 2% per year to address the renovation costs, which finally give m·X (it should be noted the difficulty in the estimated cost of renovation, so it was decided to select a conservative one) [84, 120]

Table 5.6 Technical and financial data of the equipment and infrastructure required

	LS (years)	AVAIL	η_h or η_c	c_t (€/Unit)	c_t (€/kW)	m·X (€/kW)
Transmission pipeline (heat)—Oldenburg-Wilhelmshaven	60 [123]	1.0 [124]	0.99 [125]	–	432 [108, 126]	4.3 [127]
Transmission pipeline (heat)—Bristol	60	1.0	0.99	–	144	1.4
Transmission pipeline (heat)—Cartagena	60	1.0	0.99	–	144	1.4
Heat accumulator	60 [128]	0.7 [129]	0.99 [130]	8 700 000 [131]	–	–
Centralised absorption unit for district cooling network[a]	60 [128]	0.9 [132]	0.74 [133]	–	80 [134]	3.2 [133]
Centralised gas boiler	30 [108]	0.95 [135]	0.85 [108]	–	85 [108]	3.75 [108]
District heating network—Oldenburg-Wilhelmshaven	60	1.0	0.91	5400	–	26.6
District heating network—Bristol	60	1.0	0.91	5400	–	26.6
District heating network—Cartagena	60	1.0 [136]	0.91	5400	–	26.6
District cooling network—Oldenburg-Wilhelmshaven	60	1.0	0.90	5400	–	26.6
District cooling network—Bristol	60	1.0	0.90	5400	–	26.6
District cooling network—Cartagena	60	1.0 [136]	0.91 [137]	5400 [151]	–	26.6 [138, 139]
Investment in a new wet system[b]	30 [121]	1.0	0.95 [122]	1685 [140]	–	–
Individual valve pressure	30 [121]	1.0	0.95	6657 [140]	–	–
Individual heat exchanger	30 [121]	1.0	0.95 [141]	6657 [140]	–	6.0 [142]
Individual absorption unit	15	0.9	1.00 [143]	–	496 [144]	1.7 [144]
Individual gas heater	15 [108]	0.99	0.91 [108]	–	51 [108]	3.4 [108]
Individual electric heater	15 [108]	0.99	1.00 [108]	–	220 [108]	21 [108]
Air conditioning unit	15	0.9	5.5 [145]	–	558 [145]	18.4 [145]
Individual fan coil	30 [146]	0.9	5.5	–	80 [146]	70 [147]

Source Adapted from [108, 121–150]

[a]Due to its end-use, it is assumed that the absorption chiller is a single-effect machine [148, 149]

[b]A wet system is one in which the heating pipes are directly buried in the concrete [150]

data presented in Table 5.6 has been obtained from a detailed report from Pöyry on the costs of district heating networks (see [108]).

Even for Oldenburg-Wilhelmshaven and in order to simplify the analysis, it has been supposed that the district heating network is implemented together as a whole (in this case, the worst option is assumed, which is that in which the district heating network is located as far as possible, in this case in Oldenburg).

Despite what was mentioned above, a more accurate way of addressing the analysis would be to base it on what normally happens, i.e., suppose that the district heating network consists initially of a growing number of microgrids supplied by distributed generators and that some of them are then used as peak units as the district heating network expands enough as to connect the power plants. These steps usually require about 5 years [152], so its impact on the financial feasibility of the district energy system—that, with proper maintenance, could reach 60 years [99, 153]—is limited.

Furthermore, it is assumed that all the heat accumulators of the district heating networks evaluated have a capacity of 50,000 m^3 (this amount has been chosen because it is the typical limit for atmospheric tanks [154] operating in low-temperature district heating networks [155]), allowing it, for all cases, a short-term energy storage (between 1 and 3 days) and the ability to provide services such as load following in order to integrate the maximum amount of renewable energy (mainly wind energy). Due to its higher cost and lower uptake than atmospheric tanks, it has not been contemplated the use of seasonal thermal energy storage. It should be noted that although thermal storages are usually designed for certain purposes such as peak shaving, load shifting or demand-side management, certainly in what heat accumulators are concerned (short-term energy storage, i.e. between 1 and 3 days, and load-following services), its heat capacity storage greatly varies even for similar weather conditions and demand for heat to the network (see a comparison of the heat accumulators storage for the cities of Aalborg and Odense in Appendix J).

5.3.2 Demand

The number of consumers to be supplied by the district heating network is estimated for each city through the peak heat output of the cogeneration plants in the coldest month (January) [123, 156–158], having to take into account that they should cover about 50–70% of the peak demand (the number of consumers supplied by district heating networks increases because, in addition to the heat provided by the cogeneration plants, peak demand is covered through boilers) [159, 160]; see Table 5.7 and Fig. 5.2. It was assumed that the loads related to residential buildings correspond to single-family homes (which is a fairly conservative assumption

Table 5.7 Calculation of the number of consumers (residential and nonresidential) that can be supplied during peak demand and annual demand for heating and cooling

	Oldenburg-Wilhelmshaven[a]	Bristol[b]	Cartagena[c]	Units
Rated power of the cogeneration plant (condensing mode)	757	1140	1200	MW
Thermal power cogeneration plant (backpressure mode)	931	878	923	MWt
Peak demand covered by the cogeneration plant	50%	50%	70%	
Losses during peak demand in winter	6%	6%	6%	
Demand during peak hour in the coldest month[d]	8.7	8.7	6.1	kW/dwelling
	26.2	26.2	18.5	kW other consumers
Consumers	115.1	108.5	115.0	Dwellings (thousands)
	28.4	27.1	29.0	Other consumers (thousands)
Demand for heating and hot water per unit area per year	185	160	70	$kWh/m^2/$ year (dwellings)
	250	215	95	$kWh/m^2/$ year (other consumers)
Cooling load per unit area per year	–	–	48	$kWh/m^2/$ year
Annual demand for heating and hot water	3691	3258	1411	GWh
Annual cooling demand	–	–	859	GWh

Source [165–182]

[a]Following the ratio of dwellings per inhabitant of other German cities [165], it has been estimated that the number of households for Oldenburg-Wilhelmshaven is 130,000

[b]The number of dwellings in Bristol is 184,000 [166]

[c]According to the last census of INE, Región of Murcia has 776,700 households [167] for a total of 1,476,957 inhabitants [168], so it has been estimated that the number of dwellings in Cartagena (217,641 inhabitants) [169] is 115,000 and the number of companies is 29,000

[d]To estimate the annual demand for heating, hot water, and cooling (as well as during the peak hour in the coldest month) of consumers, several publications that evaluate these parameters in cities with comparable weather conditions to the ones studied here have been consulted [170–180]. It has been considered that the surface of the dwellings is 100 m^2 [181] and the rest of the non-residential buildings (which are assumed to represent 43% of the loads for the cities studied) [182] are 220 m^2 [183]

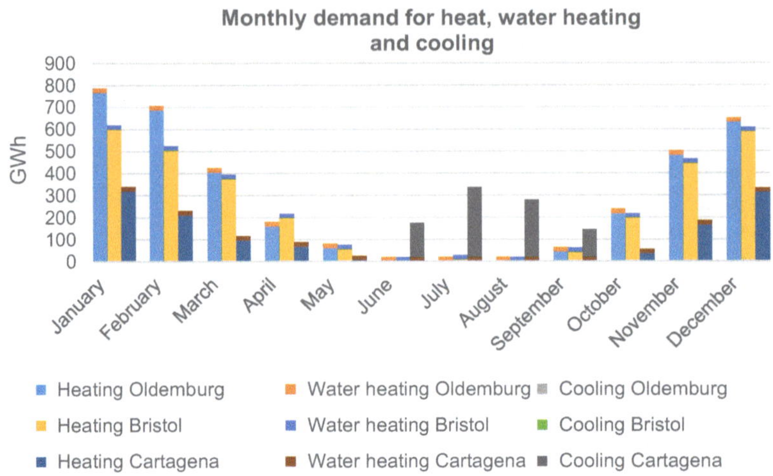

Fig. 5.2 Monthly demand for heat, water heating and cooling in Oldenburg-Wilhelmshaven, Bristol, and Cartagena. *Source* Own elaboration

because multifamily buildings would obtain even greater benefits because of their lower needs in infrastructure and fewer heat losses) [161, 162] and that those that refer to other loads are offices.

For the cost/benefit analysis, it has been considered that the losses from the cogeneration plant and loads are 9% over the year, and that heat distribution losses during peak demand in winter are 6% (this is due to the fact that, although heat losses are substantially constant throughout the year, there is a much greater amount of heat to be transmitted during peak demand in winter) [163, 164].

5.4 Results

In this section, the results of the case studies of selected cities will be presented. Taking into account that district heating networks were considered for all cities, the COGEN-a1 option represents a conventional thermal power station converted into a cogeneration plant through modifications or changes of the turbines; meanwhile, COGEN-a2 will represent a variant of COGEN-a1 in which a generating station will be built from the beginning. Due to the fact that, in comparative terms and relative to Cartagena, the cooling demand is not significant for Oldenburg-Wilhelmshaven and Bristol, only other alternatives to the current cooling for Cartagena have been considered (COGEN-b and COGEN-c cases).

Table 5.8 presents a summary of the singularities of each option.

Table 5.8 Characteristics of the options for the implementation of district heating networks in cities evaluated

Option	Characteristics
COGEN-a1	Represents a conventional power station converted into a cogeneration plant through modifications or changes of the turbines
COGEN-a2	Represents an alternative case to COGEN-a1 in which an entirely new plant will be built
COGEN-b	Represents a case in which there are district heating networks and absorption cooling units
COGEN-c	Using district heating networks and centralised cooling is assumed

5.4.1 Oldenburg-Wilhelmshaven

Tables 5.9 and K.1–K.3 (Appendix K) show the annual costs of providing heating, hot water, and cooling (plus electricity) to loads located in Oldenburg-Wilhelmshaven from one of the existing coal power plants in Wilhelmshaven (E.ON) for the three cases presented in the previous section.

The results (Table 5.10) indicate that, following the introduction of district heating networks, a reduction in the annualised cost of about €215 million would be achieved. This quantity results from subtracting the Base Case total system cost (roughly €510 million) at a discount rate of 3.5% (the one suggested by the European Commission's Directorate General Regional Policy for most member states when evaluating infrastructure investments) to the COGEN-a1 Case total system cost (roughly €295 million).

If these results were compared with those obtained for Bristol, it will be appreciated that because of a worse performance in terms of electricity production, the thermal generating station of Wilhelmshaven will have a greater annualised savings than in the case of Bristol. Meanwhile, and with regard to the overall efficiency, it will be similar both for cogeneration plants using coal as in cogeneration combined-cycle plants [184], so that the total efficiency gains will be greater in Oldenburg-Wilhelmshaven than in Bristol.

Table 5.9 Summary of results for Oldenburg-Wilhelmshaven with a discount rate of 3.5%

	Base Case	COGEN—a1	COGEN—a2	Units
Annual costs	432,710	249,049	292,623	k€
Annual heat	3691	3691	3691	GWh
Annual cooling	0	0	0	GWh
Net carbon dioxide	2570	1518	1518	kt
Net primary energy	5140	3035	3035	GWh
Net electricity	4026	3482	3482	GWh

Source Own elaboration

Table 5.10 Total system cost for each evaluated case and discount rates used in Oldenburg-Wilhelmshaven

Discount rate (%)	Base case (k€)	COGEN—a1 (k€)	COGEN—a2 (k€)
3.5	509,810	294,589	338,163
5.5	512,011	342,228	403,053
7.5	513,802	367,755	448,296

Source Own elaboration

In the case of converting the conventional coal thermal power plant of Wilhelmshaven into a cogeneration plant and implementing a district energy system, a 2.1 terawatt-hour source energy abatement is reached. Consequently, carbon footprint will be lowered in 1.05 megatonnes, and net electricity would be reduced in 0.54 terawatt-hours; some key parameters, such as key performance indicators for primary energy consumptions, are presented in Table 5.9.

The purpose of the result data presented in Table 5.9 is to summarise in a single table key information such as annual costs, heating and cooling demand, net carbon dioxide emitted to the atmosphere, net primary energy (difference between electricity purchased for heating and cooling by the households minus the electricity generated from the power plant), and net electricity (electricity purchased by the households minus electricity generated from the power plant) for the district heating schemes evaluated. It is of paramount importance to show these data as some of them are used in (4.1, 5.1–5.3), which constitute the basis of the financial assessment conducted in this book and will ultimately settle whether a given scheme is feasible from both an economic and environmental point of view. Similar remarks are also applicable to Tables 5.11 and 5.13.

5.4.1.1 Sensitivity Analysis

To study its impact on profitability, a sensitivity analysis was carried out for three different discount rates (3.5, 5.5, and 7.5%). Although the results show that the

Table 5.11 Summary of results for Oldenburg-Wilhelmshaven with a discount rate of 3.5%

	Base case	COGEN—a1	COGEN—a2	Units
Annual costs	251,935	174,663	209,941	k€
Annual heat	3258	3258	3258	GWh
Annual cooling	0	0	0	GWh
Net carbon dioxide[a]	1315	925	925	kt
Net primary energy	5480	3853	3853	GWh
Net electricity[b]	6155	5634	5634	GWh

Source Own elaboration
[a]An average cost of 30€ per tonne of carbon dioxide released has been assumed [186]
[b]See [187]

discount rate has a significant impact in absolute terms (the total system costs when "Cogen-a1" and "Cogen-a2" are evaluated at a discount rate of 7.5% are, respectively, 33 and 25% higher in comparison when a discount rate of 3.5% is used; see Table 5.10), the conversion or construction of a new cogeneration plant with the necessary infrastructure remains profitable, even with high discount rates.

5.4.2 Bristol

It is assumed that the conventional thermal power plant of Bristol, with a nominal power of 1140 MW, before the conversion, produces 6.7 TWh of electricity per year [185]. Once the plant is converted into a cogeneration plant, revenues resulting from the sale of electricity will decrease (see Tables K-4–K-6; Appendix K) due to the fact that, for the same amount of fuel used, electricity generated will decrease, for this particular case, around 10%.

Tables 5.11 and K.4–K.6 (Appendix K) show, for loads located in Bristol, the annual costs of providing heating, hot water, and cooling from the existing conventional thermal power plant (its electricity produced is also shown) for three different cases—Base Case, COGEN-a1, and COGEN-a2. Note that when the Base Case (roughly €291 million for a discount rate of 3.5%, see Table 5.12) is compared with COGEN-a1 (roughly €202 million), the annual costs of heating are reduced by €89 million due to the implementation of the district energy system. If a cogeneration station were to be constructed from the beginning (COGEN-a2 at a discount rate of 3.5% would be roughly €237 million), savings of €54 million would be achieved.

Following the introduction of low-temperature district heating networks, primary energy savings of 1.63 TWh and a reduction in CO_2 emissions of 0.39 million tonnes will take place.

5.4.2.1 Sensitivity Analysis

In order to evaluate its influence on the result, a study of how the feasibility of the system can be apportioned to different sources of uncertainty in its input (for this case, different interest rates used in discounted cash flow) was conducted. As can be seen in Table 5.12, there is a significant impact in absolute terms. For all the cases,

Table 5.12 Total system cost for each evaluated case and discount rates used in Bristol

Discount rate (%)	Base case (k€)	COGEN—a1 (k€)	COGEN—a2 (k€)
3.5	291,385	202,463	237,691
5.5	293,222	238,066	275,630
7.5	295,161	276,960	326,772

Source Own elaboration

converting a conventional thermal power plant into a cogeneration plant was advantageous; in turn, building a cogeneration plant from the beginning only was worthwhile when low interest rates in discounted cash—the ones proposed by the EU-28 for competitive regions (3.5%) and cohesion regions (5.5%)—were employed.

5.4.3 *Cartagena*

In Tables 5.13 and K-7–K-11 (Appendix K), the annual costs of providing heating, hot water, and cooling (besides electricity) from one of the existing thermal power plants in Cartagena (powerplant of El Fangal) are presented for loads located in this city.

In the Base Case (Table K-7), costs of providing heating and cooling to Cartagena for the current situation (having taken into account for this the cost of the different actual elements and the revenue from the electricity sold), have been calculated.

In the case COGEN-a1, annual costs of providing heating and cooling to the city of Cartagena, assuming the use of low-temperature district heating networks, were calculated. After evaluation, it was concluded that, in case of using this technology, annual losses would be €3 million. If a new cogeneration plant were built (COGEN-a2) (Table K-9), the annualised cost will increase €31 million.

COGEN-b assessed yearly heating and cooling expenses, assuming the use of district heating networks and absorption cooling units (see Table 5.14). The annualised savings of heating and cooling is €191.5 million.

The COGEN-c case (Table K-11) estimated the annual costs of heating and cooling in Cartagena, assuming that district energy networks (including refrigeration networks) as well as central refrigeration technologies are used. The annualised expenses reduction was €135 million.

Table 5.13 Summary of results for Cartagena with a discount rate of 3.5%

	Base case	COGEN —a1	COGEN —a2	COGEN —b	COGEN —c	Units
Annual costs	390,360	397,301	425,889	208,332	265,031	k€
Annual heat	1411	1411	1411	1411	1411	GWh
Annual cooling	859	859	859	859	859	GWh
Net carbon dioxide	1384	1237	1237	1064	1064	kt
Net primary energy	5770	5153	5153	4435	4435	GWh
Net electricity	6746	5931	5931	5931	5931	GWh

Source Own elaboration

Table 5.14 Total system cost for each evaluated case and discount rates used in Cartagena

Discount rate (%)	Base case (k€)	COGEN—a1 (k€)	COGEN—a2 (k€)	COGEN—b (k€)	COGEN—c (k€)
3.5	431,880	434,411	462,999	240,250	296,951
5.5	435,381	473,324	512,866	293,625	349,030
7.5	439,076	514,716	567,150	333,155	375,742

Source Own elaboration

All annual costs and savings related to Cogen-a1, Cogen-a2, Cogen-b, and Cogen-c are obtained by subtracting the total system costs for the Base Case at a discount rate of 3.5% from the total system costs for each of these systems. The results are presented in a quantitative manner in Table K-11.

In the case of converting the thermal power plant of El Fangal into a cogeneration plant and of using the system proposed for this case (COGEN-c), a 1.3 TWh saving is reached. Consequently, the carbon footprint would be lowered by 0.32 megatonnes and the net electricity would decrease by about 0.82 TWh (see Table K-11).

5.4.3.1 Sensitivity Analysis

In absolute terms, results are greatly influenced by discount rate changes (see Table 5.14). It can be seen that for all discount rates used, options (COGEN-b-c) are more profitable than the Base Case; conversely, no matter what discount rate is used, it is not profitable to convert the existing thermal power plant or to build a new cogeneration plant using the options (COGEN-a1-a2).

5.5 Discussion

Assessing the cost of three district heating schemes that respectively represent typical weather conditions in northern, central, and southern Europe (Oldenburg-Wilhelmshaven, Bristol and Cartagena), it has been possible to appreciate the potential reduction in expenditure as a result of both converting existing conventional thermal generating stations into cogeneration plants as well as investing in a low-temperature heating and cooling infrastructure (the latter only feasible for the weather conditions in southern Europe).

The implemented method (which considers as main elements the power plant and the needed energy to provide hot water, heating, and cooling to a specific number of loads) that sustains the economic analysis conducted for three cities of the EU-28 with different weather conditions constitutes a valuable element in itself because not using specific software and using simple and immediate calculations allows the evaluation of the potential benefits of retrofitting existing power plants into cogeneration plants with an associated heating network infrastructure (and, in the case of Cartagena, also with a cooling one), being, therefore, particularly suitable for studies with analogous purposes.

From the quantitative evidence presented in this chapter, it has been shown that the implementation of district energy systems in Oldenburg-Wilhelmshaven, Bristol, and Cartagena would reduce CO_2 emissions by 41, 30, and 23%, respectively, with the obtained net primary energy savings being similar. The savings resulting from the conversion of a conventional thermal power plant into a cogeneration one are higher in Oldenburg-Wilhelmshaven, the location with the coldest weather (and, therefore, the longer period of heating). The greatest savings are also due to the fact that, unlike the combined cycles of Bristol and Cartagena, the plant of Wilhelmshaven is a thermal coal power plant, so its electrical efficiency is lower, and its overall efficiency improvement when converted into a cogeneration plant is increased in a greater proportion.

From the evaluation of the results, some conclusions and implications for energy policy of paramount importance can be drawn. On the one hand, and for the case of Oldenburg-Wilhelmshaven, it has been shown that even with having to transport heat over long distances (more than 60 km) through heat transmission pipelines, from the joint implantation of cogeneration plants and district heating networks, significant savings and benefits from an economic and environmental perspective can be achieved. It should be noted that because district heating networks are of the low-temperature type and heat to be transmitted considerable, resulting distribution losses were lower, which has been determinant in the feasibility of all district heating schemes.

Equally striking is the case of the power plant located in Bristol. The UK, due to its population density and climatic conditions, has great potential for the implementation of district heating networks; however, and if the countries of southern Europe are excepted, diffusion is the lowest of all the EU-28. From the evaluation of the case for the power plant of Seabank, it could be seen that, like in the case of the power plant of Wilhelmshaven, great economic and environmental benefits would result from the implementation of cogeneration plants and low-temperature district heating networks.

For its part, and for the case of the power plant of El Fangal (Cartagena), it was found that although the use of district heating is not suitable for the weather conditions in Cartagena, not least is the fact that, if district heating networks and absorption refrigeration units or district heating and cooling networks were used in conjunction, economic and environmental benefits would be at the same level of other cities located in much colder climates, which gives this case a special relevance.

5.6 Conclusions

From the conducted evaluation, it was found that, in the case of deciding to convert the studied conventional thermal plants into cogeneration plants, investing in the necessary infrastructure associated means that yearly expenses for the power plants located in Oldenburg-Wilhelmshaven, Bristol, and Cartagena would decrease by 215, 89, and 192 million euros respectively, making this technology a highly attractive option from economic, energetic, and environmental perspectives (this takes into account the supposition that the discount rate is in line with the proposal by the EU Directorate-General for Regional and Urban Policy). This claim is supported by the fact that, were power plants to be converted into cogeneration plants, net electricity consumption would be reduced by about 545, 520, and 815 GWh respectively for the cases of Oldenburg-Wilhelmshaven, Bristol, and Cartagena (which represents savings of 13.5, 8.5, and 12% in comparison to the current situation). Furthermore, and in relation to the environmental benefits, CO_2 emissions would be reduced by about 1050, 390, and 320 kt respectively for the cases of Oldenburg-Wilhelmshaven, Bristol, and Cartagena (which represents savings of 40, 29, and 23% in comparison to the current situation). Consequently, this proves the attractiveness of the plan in economic, energetic, and environ mental ways.

From the evaluation conducted, it was proved that even though important, weather conditions are not determinants of the feasibility of an infrastructural project entailing the joint use of district heating networks and cogeneration, and that, taking into account the characteristics of each location, alternative solutions (such as using absorption units that utilise districts heating networks for cooling or using heating networks and integrated absorption chillers), are advantageous from an economic, environmental, and energetic perspective. This conclusion is of paramount importance as the projects evaluated in this chapter are similarly-sized district heating schemes, and consequently, the outcome of the analysis conducted in this research is certainly valuable for comparative, technological, and political purposes as it proves that weather conditions are not an impediment to the spread of such a marginally used scheme in southern Europe as district heating networks and cogeneration are. As a matter of fact, this analysis is instrumental in proving wrong the idea that the use of district heating networks and cogeneration is not an economically advantageous scheme in southern European weather conditions.

Finally, it was concluded that, in the event of using different discount rates of 3.5% (the one proposed by the Directorate General of Urban and Regional Policy of the EU), the feasibility of projects involving the joint use of cogeneration plants and district heating networks would be greatly diminished. This being the maintenance of the aforementioned discount rate of capital importance if the EU's 28 member states aim to have safe, clean, and efficient energy enacted by one of the three priorities (in particular, by the Societal Challenges) of the Horizon 2020 of the EU-28.

References

1. Mayor of London. Costs of incineration and non-incineration energy-from-waste technologies. London: Greater London Authority; 2008.
2. Friis-Jensen E. Modeling of the combined heat and power system of greater Copenhagen. Kongens Lyngby: Technical University of Denmark; 2010.
3. IEA. Energy statistics manual. Paris: OECD/IEA; 2010.
4. Tzimas V. Technology map of the European strategic energy technology plan (SET-Plan): technology descriptions. Luxembourg: European Union; 2011.
5. Khartchenko NV, Kharchenko VM. Advanced energy systems. 2nd ed. Boca Ratón: CRC Press; 2014.
6. Lako P. Combined heat and power. Paris: ETSAP; 2010.
7. OECD. The economics of long-term operation of nuclear power plants. Paris: OECD Nuclear Energy Agency; 2012.
8. Bergant R, Ploj T, Štrubelj L, Androjna G, Manojlović S, Tomažin P. Combined heat and power production in NPP Krško. Bled: 22nd International Conference Nuclear Energy for New Europe; 2013, p. 10031–9.
9. Loikala J, Itkyal S, Kaushik A, Keränen S, Nisula R, Roiha U, et al. Opportunities for Finnish environmental technology in India. Helsinki: SITRA; 2006.
10. Atkins Ltd. Revolving green fund 3 application assessments and outcomes. Bristol: HEFCE; 2013.
11. Auken S. Combined heat and power in Denmark. Copenhagen: Ministry of Environment and Energy. http://www.statensnet.dk/pligtarkiv/fremvispl?vaerkid=329&reprid=0&filid=3&iarkiv=1. Accessed Nov 2014.
12. National Research Council. Preprint, proceedings of a symposium on district heating and cooling: National Academy of Sciences Auditorium, Washington, DC, June 4–6. Washington: National Academy Press; 1984.
13. Burke A. Cartagena: Spain's largest independent power plant. Infrastruct J Proj Financ Mag. 2003;50–53.
14. Lee S. Alternative fuels. Washington: CRC Press; 1996.
15. Sallent-Cuadrado R. Return temperature influence of a district heating network on the CHP plant production costs. Gävle: University of Gävle; 2009.
16. Sysav. Heat and electricity from waste: Sysav's waste-to-energy plant. Malmö: Sysav; 2009.
17. Jonshagen K, Majed S, Magnus G. Post-combustion CO_2 capture for combined cycles utilizing hot-water absorbent regeneration. In: Proceedings of ASME turbo expo 2011, Vancouver, June 6–10; 2011.
18. Ministerio de Economía. Resolución de la Dirección General de Política Energética y Minas, por la que se autoriza a « Endesa Generación, Sociedad Anónima » , la instalación de una central térmica de ciclo combinado con cogeneración en el término municipal de Tarragona: jueves 5 diciembre 2002, BOE núm. 291. Madrid: Ministerio de Economía; 2002, p. 10358–9.
19. Pei P, Barse K, Gil AJ, Nasah J. Waste heat recovery in CO_2 compression. Int J Greenhouse Gas Control. 2014;30:86–96.
20. Lauenburg P, Wollerstrand J. Adaptive control of radiator systems for a lowest possible district heating return temperature. Energy Build. 2014;72:132–40.
21. Lončar D, Ridjan I. Medium term development prospects of cogeneration district heating systems in transition country—Croatian case. Energy. 2012;48(1):32–9.
22. Margaritis N, Rakopoulos D, Mylona E, Grammelis P. Introduction of renewable energy sources in the district heating system of Greece. Int J Sustain Energy Plan Manag. 2014;04:43–56.
23. Trygg L, Amiri S. European perspective on absorption cooling in a combined heat and power system—a case study of energy utility and industries in Sweden. Appl Energy. 2007;84:1319–37.

24. Persson U, Werner S. District heating in sequential energy supply. Appl Energy. 2012;95:123–31.
25. Sundaberg RE, Nyman HO. District heating/cogeneration application studies for the Minneapolis-St. Paul area—methods and costs estimates for converting existing buildings to hot water district heating. Oak Ridge: Oak Ridge National Laboratory; 1979.
26. Sundaberg RE, Leas R, Frost E, Hagstrom K. District heating/cogeneration application studies for the Minneapolis-St. Paul area: institutional issues of a new district heating/cogeneration system—ownership options, barriers, and implementation strategy. Oak Ridge: Oak Ridge National Laboratory; 1980.
27. Barnes MH, Abrahamsson B. District heating/cogeneration application studies for the Minneapolis-St. Paul area—market assessment and economic analysis of the St. Paul district heating system (Vol. II). Oak Ridge: Oak Ridge National Laboratory; 1983.
28. Margen P, Cronholm LA, Larsson K, Marklund JE. District heating/cogeneration application studies for the Minneapolis-St. Paul area—Executive summary: overall feasibility and economic viability for a district heating/new cogeneration system in Minneapolis, St. Paul. Oak Ridge: Oak Ridge National Laboratory; 1979.
29. Kolb JO, Bauman HF, Jones PD. Technical feasibility and economics of retrofitting an existing nuclear power plant to cogeneration for hot water district heating. Oak Ridge: Oak Ridge National Laboratory; 1984.
30. Olszewski M. Power plant rejected heat utilization: an assessment of the potential for wide-scale implementation. Oak Ridge: Oak Ridge National Laboratory; 1977.
31. Nyman HO, Kolb JO, Krautbauer M. District heating/cogeneration application studies for the Minneapolis-St. Paul area—market assessment and economic analysis of the St. Paul district heating system. Oak Ridge: Oak Ridge National Laboratory; 1983.
32. Englesson GA, Pavlenco GF, Hodson JS, Lee NH, García AN, Menaker BJ. District heating/cogeneration application studies for the Minneapolis-St. Paul area—economic comparison of new coal-fueled, cogeneration power plants for district heating and electricity-only and heat-only power plants. Oak Ridge: Oak Ridge National Laboratory; 1982.
33. Englesson GA, Casapis MC, Pavlenco GF, Menaker B, Lee NH, Denesdi L. District heating/cogeneration application studies for the Minneapolis-St. Paul area—modifications of the existing units at the high bridge power plant to cogeneration for hot water district heating. Oak Ridge: Oak Ridge National Laboratory; 1980.
34. Lund H. District heating and market economy in Latvia. Energy. 1999;24(7):549–59.
35. Lund H, Möller B, Mathiesen BV, Dyrelund A. The role of district heating in future renewable energy systems. Energy. 2010;35(3):1381–90.
36. Nielsen S, Möller B. GIS based analysis of future district heating potential in Denmark. Energy. 2013;57:458–68.
37. Dotzauer E. Experiences in mid-term planning of district heating systems. Energy. 2003;28 (15):1545–55.
38. Rezaie B, Rosen MA. District heating and cooling: review of technology and potential enhancements. Appl Energy. 2012;93:2–10.
39. Chow TT, Au WH, Yau R, Cheng V, Chan A, Fong KF. Applying district-cooling technology in Hong Kong. Appl Energy. 2004;79:275–89.
40. Piacentino A, Barbaro C, Cardona F, Gallea R, Cardona E. A comprehensive tool for efficient design and operation of polygeneration-based energy lgrids serving a cluster of buildings. Part I: description of the method. Appl Energy. 2013;111:1204–21.
41. Pusat S, Erdem HH. Techno-economic model for district heating systems. Energy Build. 2014;72:177–85.
42. Möller B, Lund H. Conversion of individual natural gas to district heating: geographical studies of supply costs and consequences for the Danish energy system. Appl Energy. 2010;87:1846–57.

43. Eriksson O, Finnveden G, Ekvall T, Björklund A. Life cycle assessment of fuels for district heating: a comparison of waste incineration, biomass, and natural gas combustion. Energy Policy. 2007;35:1346–62.
44. Nitkiewicz A, Sekret R. Comparison of LCA results of low temperature heat plant using electric heat pump, absorption heat pump and gas-fired boiler. Energy Convers Manag. 2014;87:647–52.
45. Nitkiewicz A, Sekret R. Comparison of district heating CHP and distributed generation CHP with energy, environmental and economic criteria for Northern Italy. Energy Convers Manag. 2015;92:114–28.
46. Finney KN, Chen Q, Sharifi VN, Swithenbank J, Nolan A, White S. Developments to an existing city-wide district energy network: part II—analysis of environmental and economic impacts. Energy Convers Manag. 2012;62:176–84.
47. Ghafghazi S, Sowlati T, Sokhansanj S, Melin S. A multicriteria approach to evaluate district heating system options. Appl Energy. 2010;87:1134–40.
48. Tana S, Hashima H, Lee C, Taib MR, Yan J. Economical and environmental impact of waste-to-energy (WTE) alternatives for waste incineration, landfill and anaerobic digestion. Energy Proced. 2014;61:704–8.
49. Udomsri S, Martin AR, Martin V. Thermally driven cooling coupled with municipal solid waste-fired power plant: application of combined heat, cooling and power in tropical urban areas. Appl Energy. 2011;88:1532–42.
50. Åberg M. Investigating the impact of heat demand reductions on Swedish district heating production using a set of typical system models. Appl Energy. 2014;118:246–57.
51. Wahlund B, Yan J, Westermark M. A total energy system of fuel upgrading by drying biomass feedstock for cogeneration: a case study of Skellefteå bioenergy combine. Biomass Bioenerg. 2002;23:271–81.
52. Song H, Starfelt F, Daianova L, Yan J. Influence of drying process on the biomass-based polygeneration system of bioethanol, power and heat. Appl Energy. 2012;90:32–7.
53. Li H, Wang W, Yan J, Dahlquist E. Economic assessment of the mobilized thermal energy storage (M-TES) system for distributed heat supply. Appl Energy. 2013;104:178–86.
54. Wahlund B, Yan J, Westermark M. Increasing biomass utilisation in energy systems: a comparative study of CO_2 reduction and cost for different bioenergy processing options. Biomass Bioenerg. 2004;26:531–44.
55. Li H, Sun Q, Zhang Q. Wallin. A review of the pricing mechanisms for district heating systems. Renew Sustain Energy Rev. 2015;42:56–65.
56. Ma Z, Li H, Sun Q, Wang C, Yan A, Starfelt F. Statistical analysis of energy consumption patterns on the heatdemand of buildings in district heating systems. Energy Build. 2014;85:464–72.
57. Ristimäki M, Säynäjoki A, Heinonen J, Junnila S. Combining life cycle costing and life cycle assessment for an analysis of a new residential district energy system design. Energy. 2013;63:168–79.
58. Oliver-Solá J, Gabarrell X, Rieradevall J. Environmental impacts of the infrastructure for district heating in urban neighbourhoods. Energy Policy. 2009;37:4711–9.
59. Yu D, Tan H, Ruan Y. An improved two-step floating catchment area method for supporting district building energy planning: a case study of Yongding County city, China. Appl Energy. 2012;95:156–63.
60. Streckiene G, Martinaitis V, Andersen AN, Katz J. Feasibility of CHP-plants with thermal stores in the German spot market. Appl Energy. 2009;86:2308–16.
61. Krajacic G, Duić N, Zmijarević Z, Mathiesen BV, Vučinić AA, Carvalho MG. Planning for a 100% independent energy system based on smart energy storage for integration of renewables and CO_2 emissions reduction. Appl Therm Eng. 2011;31(13):2073–83.
62. Jožef Stefan Institute. Methodology for determining the reference costs for high-efficiency cogeneration. Ljubljana: Ministry of the Economy of the Republic of Slovenia; 2009.
63. Anderson D. Electricity generation costs and investment decisions: a review. London: Imperial College Centre for Energy Policy and Technology; 2007.

64. Newell SA, Hagerty JM, Spees K, Pfeifenberger JP, Liao Q, Ungate CD, et al. Cost of new entry estimates for combustion turbine and combined cycle plants in PJM. Cambridge: The Brattle Group; 2014.
65. Vuorinen A. National power system planning. Wärtsilä Tech J. 2007;02(2007):21–5.
66. Paulson CAJ, Durie RA, McMullan P, Smith AY, Williams DJ. Greenhouse gas control technologies: proceedings of the 5th international conference on greenhouse gas control technologies. Collingwood: CSIRO; 2001.
67. Dimian AC, Bildea C, Kiss A. Integrated desing and simulation of chemical processes. Radarweg: Elsevier; 2014.
68. Levidow L, Papaioannou T, Borda-Rodríguez A. Innovation priorities for UK bioenergy: technological expectations within path dependence. Sci Technol Stud. 2013;26(3):14–36.
69. Policy Directorate General Regional. Guide to cost benefit analysis of investment projects. 1st ed. Brussels: European Commission; 2008.
70. Florio M, Maffii S, Atkinson G, De Rus G, Evans D, Ponti M, et al. Structural funds, cohesion fund and instrument for pre-accession. Brussels: European Commission; 2008.
71. Cambridge Economic Policy Associates and Ricardo—AEA. Development of phase II of the Northern Ireland Renewable Heat Incentive. Belfast: Department of Enterprise, Trade and Investment; 2013.
72. Element Energy Limited and AEA Group. 2050 options for decarbonising heat in buildings: committee on climate change—final report. Cambridge: Element Energy Limited; 2012.
73. Ove Arup & Partners. Stockport town centre district energy system: recommended actions to create phase 1. Stockport: Stockport Metropolitan Borough Council; 2011.
74. Ricardo-AEA. The durability of products: task 1 report. Standard assessment for the circular economy under the eco-innovation action plan. Brussels: European Commission; 2014.
75. Economic Regulation Authority. Floor and ceiling costs to apply to the public transport authority. Perth: Government of Western Australia; 2004.
76. De Jong P. Economic, technical and environmental analysis of renewable and non-renewable electricity generation technologies in Brazil. Salvador de Bahía, Federal University of Bahia; 2013.
77. Simoes S, Nijs W, Ruiz P, Sgobbi A, Radu D, Bolat P, et al. The JRC-EU-TIMES model—assessing the long-term role of the SET plan energy technologies. Luxembourg: European Commission; 2013.
78. Müller B, Michaelowa A, Vrolijk C. Rejecting Kyoto: a study of proposed alternatives to the Kyoto Protocol. London: Climate Strategies; 2001.
79. DNV Climate Change Services. Third party assessment of the comprehensive refurbishment of the Prunéřov II power plant. Prague: Ministry of the Environment of the Czech Republic; 2010.
80. Bauer N, Brecha RJ, Luderer G. Innovation priorities for UK bioenergy: technological expectations within path dependence. Proc Natl Acad Sci. 2012;109(42):16805–10.
81. Lahmeyer International. Energy interconnection Europe—Malta: Final report—work package IIA; 2008.
82. Punnonen K. Small and Medium size LNG for Power Production. Vaasa: Wärtsilä Finland Oy; 2013.
83. IEA. World energy investment outlook. Insights. Paris: OECD; 2003. p. 2003.
84. Larsson S. Reviewing electricity generation cost assessments. Uppsala: Uppsala Universitet; 2012.
85. Rosnes O, Vennemo H. Powering up: costing power infrastructure spending needs in Sub-Saharan Africa. Washington: The World Bank; 2009.
86. IAEA. Comparative studies of energy supply options in Poland for 1997–2020. Vienna: International Atomic Energy Agency; 2002.
87. Bureau of Resources and Energy. Economics Australian energy technology assessment 2013 model update. Canberra: Commonwealth of Australia; 2013.
88. Eurostat. Gas prices for domestic consumers—bi-annual data. Luxembourg: Eurostat. http://ec.europa.eu/eurostat/en/web/products-datasets/-/NRG_PC_202. Accessed 3 Oct 2015.

89. Union European. EU Energy, transport and GHG emissions trends to 2050: reference scenario 2013. Luxembourg: European Union; 2014.
90. Market Observatory for Energy. Quaterly report on European electricity markets: volume 6, issue 2. Brussels: European Commission; 2013.
91. Eurostat. Gas prices for industrial consumers, from 2007 onwards—bi-annual data. Luxembourg: Eurostat; 2014. http://appsso.eurostat.ec.europa.eu/nui/submitViewTableAction.do. Accessed 10 Jan 2015.
92. European Union. Electricity prices for domestic consumers, from 2007 onwards–bi-annual data; 2014. http://appsso.eurostat.ec.europa.eu/nui/show.do?dataset=nrg_pc_204&lang=en. Accessed 12 Jan 2015.
93. IEA. Electricity information 2012. Paris: OECD/IEA; 2012.
94. Vikkelso A, Larsen JHM, Sørensen HC. The Middelgrunden offshore wind farm: a popular initiative. Copenhagen: Copenhagen Environment and Energy Office; 2003.
95. MWH. Electricity emission factor review. London: European Bank for reconstruction and development; 2009.
96. Osman AE. Life cycle optimization model for integrated cogeneration and energy systems applications in buildings. Pittsburgh: University of Pittsburgh; 2006.
97. Sánchez-Castaño J. Analysis of a new district heating line evaluation of heat losses and hydraulic facilities. Gävle: University of Gävle; 2008.
98. SEAI. Derivation of primary energy and CO_2 factors for electricity. Dublin: Sustainable Energy Authority of Ireland; 2010.
99. Energinet. Technology data for energy plants: generation of electricity and district heating, energy storage and energy carrier generation and conversion. Erritsø: Energinet; 2012.
100. Eurelectric. Efficiency in electricity generation. Brussels: Union of the Electricity Industry; 2003.
101. Egenhofer C, Jansen JC, Bakker SJA, Jussila-Hammes J. Revisiting EU policy options for tackling climate change: a social cost-benefit analysis of GHC emissions reduction strategies. Brussels: Centre for European Policy Studies; 2006.
102. European Commission. How to develop a Sustainable Energy Action Plan (SEAP)—Guidebook. Luxembourg: Publications office of the European Union; 2010.
103. Hill N, Venfield H, Dun C, James K. 2013 Government GHC conversion factors for company reporting: methodology paper for emission factors. London: Department for Environment Food & Rural Affairs; 2013.
104. AECOM. Smart city—intelligent energy integration for London's decentralised energy projects. London: Greater London Authority; 2012.
105. Amiri S. Economic and environmental benefits of CHP-based district heating systems in Sweden. Linköping: Linköping Institute of Technology; 2013.
106. Beer M, Huber M, Mauch W. Flexible operation of cogeneration plants—chances for the integration of renewables. Munich: The Research Center for Energy Economics; 2012.
107. Molyneaux A, Leyland G, Favrat D. Environomic multi-objective optimisation of a district heating network considering centralized and decentralized heat pumps. Energy. 2010;35 (2):751–8.
108. Pöyry. The potential and costs of district heating networks. Oxford: Pöyry Energy Ltd; 2009.
109. RWE, 2005. Frimmersdorf and Neurath power plants: electricity from Rhenish lignite. Bergheim: RWE Power AG; 2005.
110. Gargiulo M. Getting started with TIMES-VEDA: version 2,7. Paris: IEA—Energy Technology Systems Analysis Programme; 2009.
111. Harvey D. A handbook on low-energy buildings and district-energy systems: fundamentals, techniques and examples. New York: Taylor & Francis; 2006.
112. Gadd H, Werner S. Achieving low return temperatures from district heating substations. Appl Energy. 2014;136:59–67.
113. Rydstrand M. An analysis of the efficiency and economy of humidified gas turbines in district heating applications. Energy. 2004;29(12–15):1945–61.

114. Danfoss A/S. The heating book—8 steps to control of heating systems. Nordborg: Danfoss A/S; 2008.
115. Harvey DA. Clean building: contribution from cogeneration, trigeneration and district energy. Waltham Abbey: Cogeneration and on-site power production (September–October 2006); 2006, p. 107–15.
116. Breeze P. The cost of power generation: the current and future competitiveness of renewable and traditional technologies. Warwick: Business Insights; 2010.
117. House of Lords. Energy efficiency—volume II: evidence (2nd report of Session 2005–06). London: House of Lords; 2005.
118. Davison J. Retrofitting CO_2 capture to existing power plants: report 2011/02. Cheltenham: IEAGHC; 2011.
119. Parsons Brinckerhoff. Thermal power station advice—fixed & variable O&M costs: report for the electricity commission. Wellington: Electricity Commission; 2009.
120. Blyth W. The economics of transition in the power sector. Paris: International Energy Agency; 2010.
121. Euroheat & Power. District heating in buildings. Brussels: Euroheat & Power; 2011.
122. Rhodes M. A new model for ECO; London: Encraft Ltd; 2014. www.encraft.co.uk/viewpoints/a-new-model-for-eco. Accessed 28 Aug 2014.
123. Dalla Rosa A, Boulter R, Church K, Svendsen S. District heating (DH) network design and operation toward a system-wide methodology for optimizing renewable energy solutions (SMORES) in Canada: a case study. Energy 2012;45(1):960–74.
124. Nuorkivi A. To the rehabilitation strategy of district heating in economies in transition. Espoo: Helsinki University of Technology; 2005.
125. Zhang L, Lianzhong L, Oddgeir G, Jan Eric T, Hongwei L, Svend S. China's Anshan project—a good example to implementing Scandinavian technology and environmentally friendly heat source to upgrade the DH system. Stockholm: The 14th International Symposium on District Heating and Cooling; 2014.
126. Riddle A. District heating & the future smart cities approach. Copenhagen: Ramboll Group A/S; 2013.
127. Kalkumn B. Improvement of district heating in Kosovo: final report. Heidelberg: Energy & Utility Consulting; 2009.
128. Ånestad A. Net electricity load profiles of zero emission buildings: a cost optimization investment model for investigating zero balances, operational strategies and grid restrictions. Trondheim: Norwegian University of Science and Technology; 2014.
129. Gil S, Gill S, Dolan MJ, Frame D, Ault GW. The role of electric heating and district heating networks in the integration of wind energy to island networks. Int J Distrib Energy Resour. 2010;7(3):245–63.
130. Akkaya BM, Romanchenko D. Modeling and analysis of a district heating system containing thermal storage: case study of the district heating of Borås. Göteborg: Chalmers University of Technology; 2013.
131. Kjær Petersen M, Aagaard J. Heat accumulators: news from DBDH 1/2004. Copenhagen: Energi E2; 2004.
132. Calixto E. CENPES II project reliability analysis: safety and reliability for managing risk. Estoril: European Safety and Reliability Conference 2006; 2006. p. 2461–2468.
133. Capital Cooling, 2014. District cooling and the customers' alternative cost: work package 2. Dresden: Renewable Smart Cooling for Urban Europe.
134. Goodheart KA. Low firing temperature absorption chiller system. Madison: University of Wisconsin; 2000.
135. Holmberg J. Our dream—a city free from fossil fuels: production of district heating, district cooling, electricity and biogas. Borås: Borås Energi och Miljö; 2009.
136. Kaan H. District heating systems with low losses. Petten: Demohouse; 2008.
137. Kankaro B. Smart district heating from Finland: the coldest country in Europe, the best country of district heating. Helsinki: Cleantech Finland; 2013.

138. Department of Energy & Climate Change. European energy efficiency: analysis of ODYSSEE indicators. London: Department of Energy & Climate Change; 2012.
139. Ehrig R, Kristöfel C, Pointner C. Operating figures, quality parameters and investment costs for district heating systems. Wieselburg-Land: Bioenergy 2020+; 2011. http://www.afo.eu.com/default.asp?SivuID=28291. Accessed 27 Dec 2014.
140. Rafferty K. Selected cost considerations for geothermal district heating in existing single-family residential areas. GHC Bull. 1996. p. 10–5.
141. Kopuničová M. Feasibility study of binary geothermal power plants in Eastern Slovakia: analysis of OCR and Kalina power plants. Akureyri: The School for Renewable Energy Science; 2009.
142. Eggen G, Vangsnes G. Heat pump for district cooling and heating at Oslo Airport, Gardermoen. Trondheim: Sintef; 2004.
143. Pérez de Viñaspre M, Bourouis M, Coronas A, García A, Soto V, Pinazo JM. Monitoring and analysis of an absorption air-conditioning system. Energy Build. 2004;36(9):933–43.
144. Yakazi Europe Limited. Concepts for energy efficient cooling: absorption chiller installation in a hotel; 2008. http://www.yazaki-airconditioning.com/applications/case_studies/hotel.html. Accessed 30 Dec 2014.
145. Northwest Power & Conservation Council. Regional costs and bulk power system benefits. Portland: Northwest Power & Conservation Council; 2000. http://www.nwcouncil.org/media/6665/EStarHPandACUpgradeMHPTCSrev.xls. Accessed 1 Jan 2015.
146. Dinçer I, Zamfirescu C. Sustainable energy systems and applications. New York: Springer; 2011.
147. Edwards Valance. Valance vs. fan coils units. Pompton Plains: Chiller Solutions LLC; 2008.
148. Keil C. Customized absorption heat pumps for utilization of low-grade heat sources. Garching: Bavarian Center for Applied Energy Research; 2008.
149. Herold KE, Radermacher R, Klein SA. Absorption chillers and heat pumps. Boca Ratón: CRC Press; 1996.
150. Nowak S. Reducing energy use of an electric floor heating system and analyzing thermal comfort and heat transmission when using different control strategies. Gävle: University of Gävle; 2014.
151. Grohnheit PE. Socio economic research on fusion EFDA technology workprogramme 2011 WP11-SER-ETM-1,3,6: Activity 1,3,6. modelling infrastructures—final report. Roskilde: Technical University of Denmark; 2012.
152. Schmidt RR, Fevrier N, Dumas P. Smart cities: stakeholder platform—smart thermal grids. 2nd Version. Brussels: European Commission; 2013.
153. Dalla Rosa A. Toward 4th generation district heating: experience and potential of low-temperature district heating—Annex X final report. Paris: IEA; 2014.
154. DBDH. DHC in Denmark: characteristics. Frederiksberg: Danish Board of District Heating; 2014. http://dbdh.dk/characteristics/. Accessed 18 May 2015.
155. Ramboll. Thermal storage in district heating systems. Copenhagen: Ramboll; 2014. http://www.districtenergy.org/assets/pdfs/03AnnualConference/Monday-A/A4,1OVERBYERamboll-A4ThermalStoresinDH-slidesonly.pdf. Accessed 18 May 2015.
156. Pirouti M, Bagdanavicius A, Ekanayake J, Wu J, Jenkins N. Energy consumption and economic analyses of a district heating network. Energy. 2013;57:149–59.
157. Torchio MF, Genon G, Poggio A, Poggio M. Merging of energy and environmental analyses for district heating systems. Energy. 2009;34(3):220–7.
158. Dalla Rosa A, Christensen JE. Low-energy district heating in energy-efficient building areas. Energy 2011;36(12):6890–99.
159. Energy Saving Trust. The applicability of district heating for new dwellings. London: Energy Saving Trust; 2008.
160. Ziębik A, Gładysz P. Optimal coefficient of the share of cogeneration in district heating systems. Energy. 2012;45(1):220–7.
161. Brage A. District heating form biomass in Borensberg. Stockholm: Swedish Environmental Protection Agency; 2010.

162. Gustavsson L, Karlsson Å. Heating detached houses in urban areas. Energy. 2003;28 (8):851–75.
163. Jones C. Utilising nuclear energy for low carbon heating services in the UK. Manchester: University of Manchester; 2013.
164. Li H, Svendsen S. Energy and exergy analysis of low temperature district heating network. Energy. 2012;45(1):237–46.
165. Britannica Encyclopaedia. Britannica book of the year (2008). Chicago: Encyclopaedia Britannica Inc.; 2008.
166. Department for Communities and Local Government. Number of households, 2008 mid-year estimate, Bristol, City of UA; 2014. http://opendatacommunities.org/showcase/dashboard/local_authorities/unitary-authority/bristol. Accessed 5 Dec 2014.
167. INE. Censos de población y viviendas 2011: Edificios y viviendas. Madrid: Instituto Nacional de Estadística; 2013.
168. INE. Banco de series temporales; 2014. http://www.ine.es/consul/serie.do?d=true&s=EPOB23940&nult=15. Accessed 4 Dec 2014.
169. INE. Nomenclator: Población del padrón continuo por unidad poblacional; 2015. http://www.ine.es/nomen2/index.do?accion=busquedaAvanzada&entidad_amb=no&codProv=30&codMuni=016&codES=0&codNUC=&L=&ano=2011. Accessed 2 Jan 2015.
170. Zangheri P, Armani R, Pietrobon M, Pagliano L, Fernandez Boneta M, Müller A. Heating and cooling energy demand and loads for building types in different countries of the EU: D2,3. of WP2 of the Entranze Project. Vienna: Entranze Project; 2014.
171. Biaou A, Bernier M. Achieving total domestic hot water production with renewable energy. Build Environ. 2008;43(4):651–60.
172. Van Dijk D, Spiekman M, de Wilde P. Monthly method to calculate cooling demand for EP regulations (progress report). Delft: TNO Building and Construction Research; 2004.
173. European Energy Agency. Energy efficiency and energy consumption in the household sector (ENER 022)—Assessment published Apr 2012. Copenhagen: European Energy Agency; 2012. http://www.eea.europa.eu/data-and-maps/indicators/energy-efficiency-and-energy-consumption-5/assessment. Accessed 5 Jan 2014.
174. Tacis. Optimisation of energy supply and demand in municipalities: the example of Tver. Brussels: European Commission; 1995.
175. Manyes A, Sisó L, Salom J. Block level study and simulation for residential retrofitting. Chambéry. In 13th conference of international building performance simulation association; 2013, p. 88–95.
176. Guarino F, Salom J, Cellura M. Modeling of Spanish household electrical consumptions: Simplified and detailed stochastic approach in TRNSYS environment. Chambéry. In: 13th conference of international building performance simulation association; 2013. p. 2436–43.
177. Rees M. The integrated design of new build multi vector energy supply schemes. Cardiff: Cardiff University; 2012.
178. IDOM. Análisis de contexto de la Región de Murcia (Multisectorial): Gestión de contenidos "Foros de Innovación"—Programa InnoCámaras. Murcia: InnoCámaras; 2010.
179. Department for Business Innovation & Skills. Business population estimates for the UK and regions 2013. London: Department for Business Innovation & Skills; 2013.
180. Smith DW. Cold regions utilities monograph. 3rd ed. Reston: American Society of Civil Engineers; 1996.
181. McKinnon D, Bolland T, Kallay TK, Bakas I, Buttler M, Von Geibler J, et al. Housing assessment: final report ETC/SCP 2012, Task 2.5.1.1. Copenhagen: European Topic Centre on Sustainable Consumption and Production; 2013.
182. Menkveld M, Beurskens L. Renewable heating and cooling in the Netherlands: D3 of WP2 from the RES-H policy project. Petten: Energy Research Centre of the Netherlands; 2009.
183. Economidou M. Europe's building under the microscope: a country-by-country review of the energy performance of buildings. Brussels: Buildings Performance Institute Europe; 2011.

184. Laitner JA. Energy efficiency investments as an economic productivity strategy for Texas. Washington: American Council for an Energy-Efficient Economy; 2011.
185. Morbee J. Analysis of energy saving potentials in energy generation: Final results. Luxembourg: European Union; 2012.
186. Turner A. Meeting carbon budgets—the need for a step change. London: Parliament Committee on Climate Change; 2009.
187. Manicuta M. Aspects on the development of the regulatory framework for promoting renewable energy in Romania. Bucarest: Autoritatea Naţională de Reglementare în domeniul Energiei; 2010.

Appendix A
Sankey Diagram with the EU Energy Balance for 2012

See Fig. A.1.

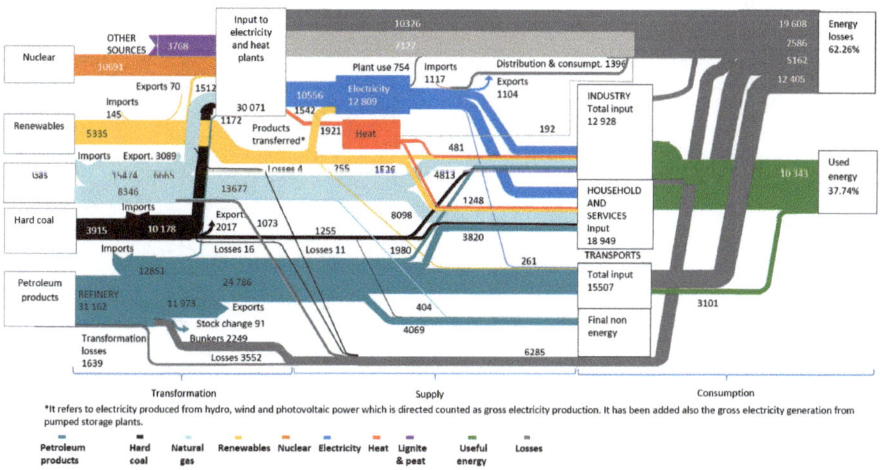

Fig. A.1 Sankey diagram with the EU energy balance for 2012. *Source* Adapted from [1]

A. Colmenar-Santos et al., *District Heating and Cooling Networks in the European Union*, DOI 10.1007/978-3-319-57952-8

Appendix B
Historical of District Heating Prices, Market Share

In (Fig. B.1), the development of district heating prices for different European countries is presented from 2008 to 2012, while (Fig. B.2) does the same with the percentage of users who have access to district heating networks for different countries that belong to the OECD, being of a 12.4% in 2012 for the member countries of the European Union. Finally, the contribution of cogeneration to the production of electricity for some OECD countries is presented in (Fig. B.3).

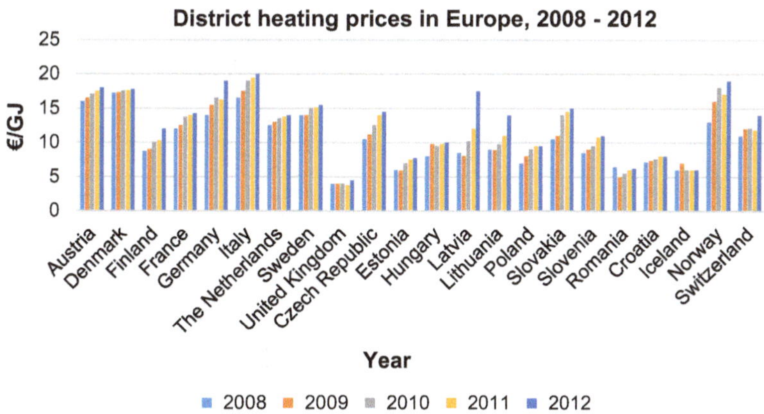

Fig. B.1 Price of heat supplied by district heating networks placed in Europe for the period 2008–2012 [2]

A. Colmenar-Santos et al., *District Heating and Cooling Networks in the European Union*, DOI 10.1007/978-3-319-57952-8

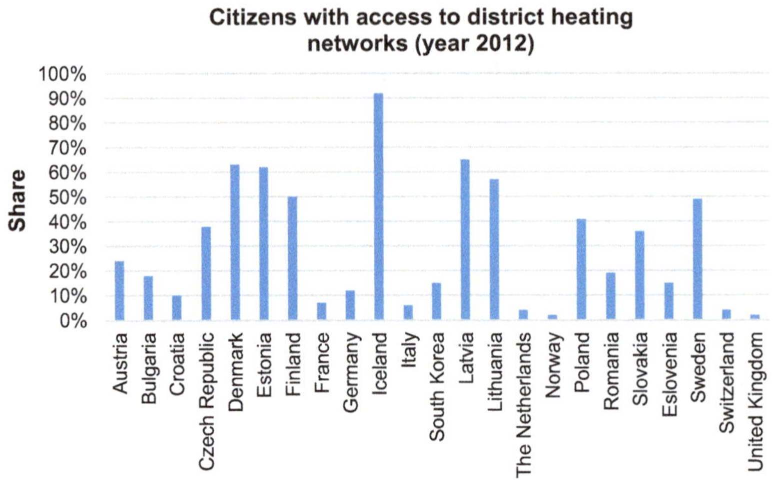

Fig. B.2 Percentage of citizens who have access to district heating networks for different countries of the OECD (2012) [2]

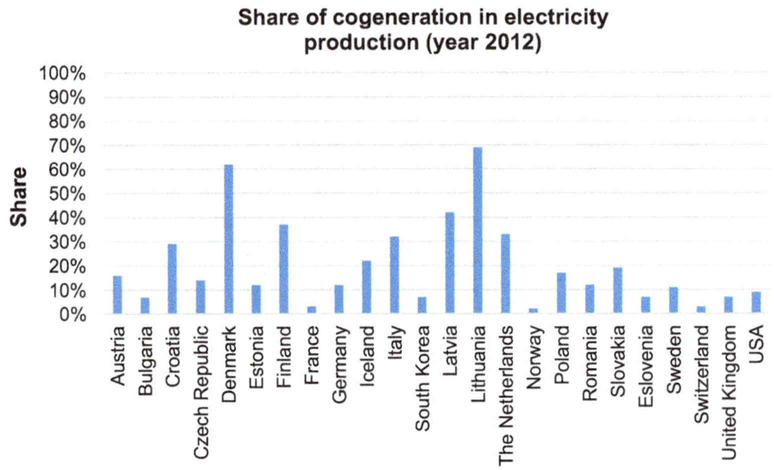

Fig. B.3 Contribution of CHP in electricity production for different countries of the OECD (2012) [2]

Appendix C
Large Scale Heat Transport Technology

Heat transmission expenses through large-diameter pipes (even with distances being considerable) is low [3–5]. As an example, [3] shows that the cost of transmitting 2 GW of heat through large-diameter pipes at a distance of about 140 km would be about 0.25 €/kWh, this being due to the fact that the ability to transport heat through the pipe is proportional to the square of the diameter, while the cost of the pipe increases proportionally with the diameter [3]. Therefore, the larger the piping and the demands of users, the greater the heat loss and the cost per unit of energy delivered (Table C.1).

An example of large-scale heat transport technology can be cited the case of Aarhus or that of Prague. Although the distance from the cogeneration plant to the center of the city of Aarhus is only 20 km [7], the total length of the heat transport network rises to 130 km [8], representing the total length of all pipelines that are

Table C.1 Comparison of the heat transport cost depending on the diameter and length of the pipe [6]

DN	Flow	Capacity	Price pr. km trench	Price pr. km pr. capacity	Price pr. km pr. annual sale	Heat loss pr. km
mm	m/s	MW	m€/km	€/km/MW	€/km/MWh/year	%/km
100	1.0	2	0.6	274,000	68.0	2.96
200	1.5	13	1.1	78,000	20.0	1.08
300	2.0	38	1.7	45,000	11.0	0.50
400	2.3	69	2.2	31,300	7.0	0.28
500	2.6	125	2.8	22,400	6.0	0.15
600	2.9	203	3.4	16,700	3.5	0.11
700	3.2	301	3.9	12,600	3.5	0.09
800	3.5	431	4.3	10,000	2.5	0.07
900	3.5	551	4.9	9000	2.5	0.06
1000	3.5	681	5.7	8200	2.5	0.05
		Supply temperature (pre-insulated pipes)				120 °C
		Return temperature (designed maximum operating temperature = 130 °C)				60 °C
		Pressure loss (variable flow pumps)				10 mm/m

A. Colmenar-Santos et al., *District Heating and Cooling Networks in the European Union*, DOI 10.1007/978-3-319-57952-8

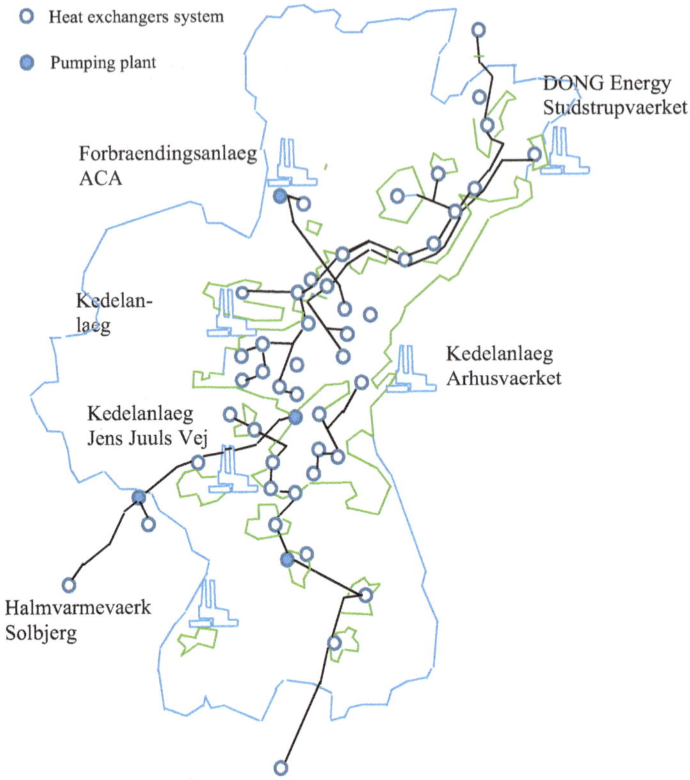

Fig. C.1 Heat transport system of Aarhus [9]

continuously connected. In (Fig. C.1), it can be observed that this large-scale heat transport network is part of a system consisting of generating plants (Studstrupvaerket), closed vessels in which water or other fluid is heated for consumption peaks, and apparatuses for burning waste material (ACA).

Fig. C.2 Pipes for transporting large-scale heat (2xDN 1200) between Melnik and Prague [11]

As for Prague, heat transport piping from the cogeneration plant of Melnik have a diameter of 1.2 m and a system length of 63.5 km [10]. As can be seen in (Fig. C.2), to cover much of the distance between Melnik and Prague, pipes are laid above the surface.

Appendix D
Heat Distribution in Modern District Heating Systems

In modern systems, the heat distribution is almost always conducted through the use of pre-insulated pipes that distribute hot water [12] (HWP are usually buried directly in the ground, so their installation is cheap, requiring no maintenance) [13].

Furthermore, it is noteworthy that to obtain a higher efficiency regarding electricity, heat temperature extraction should be as small as possible, being as consequences that the heat is cheaper in fuel costs and the carbon dioxide emissions are lower [14]. These are the main reasons why there is a trend towards the use of water at a lower temperature—although, in a limited number of systems, such a small temperature as 45–55 °C may be working properly [15], a temperature between 70 and 80 °C is a more "reasonable" one [16].

In some circumstances, and very especially in locations with flat terrain where district heating systems have a low-temperature design (below 90 °C) [17–20], a direct connection from the dwelling to the principal district heating network can be performed [18]. This approach has the advantage of the ability to use lower temperatures and provide the highest efficiency possible, thus giving lower losses per unit of heat used. With a direct connection, it is possible to eliminate the cost of any heat exchanger [18]. A valve that reduces the pressure for each consumer to prevent from overpressure to end-user facilities is used [18]. Direct connections, therefore, have many advantages because it is not necessary to allocate any investment for heat exchangers and it allows a more efficient operation of cogeneration regarding energy generation [18].

A. Colmenar-Santos et al., *District Heating and Cooling Networks in the European Union*, DOI 10.1007/978-3-319-57952-8

Appendix E
Thermal Storage in Accumulators for District Heating Systems

Currently, a very limited percentage of the EU-28 produced electricity is stored (predominantly in reversible hydroelectric power stations) [21, 22], so any significant increase (such as by the use of wind energy for heating buildings) would be extremely costly [23, 24] and would involve a considerable increase in the capacity of transmission and distribution [25]. The electricity storage through isentropic heat pumps is currently the only long-term electricity technology storage that can be used in the electrical network in an economical way, having a cost of about 35 €/kWh [26]. Meanwhile, the thermal storage is considerably cheaper as the cost varies in a band range of 0.1–10 €/kWh, in this case not needing any upgrading of the electricity grid [27].

Thermal storage allows the cogeneration plant to generate electricity whenever and to increase system flexibility due to the fact that thermal storage heat output change is almost immediate and that, unlike backpressure cogeneration plants that have to generate heat even when electricity price is high, condensing-extraction cogeneration plants have increased flexibility thanks to the possibility of generating electricity exclusively [28, 29]. In (Figs. E.1 and E.2) are respectively shown the heat storage directly and indirectly connected to the district heating network.

A. Colmenar-Santos et al., *District Heating and Cooling Networks in the European Union*, DOI 10.1007/978-3-319-57952-8

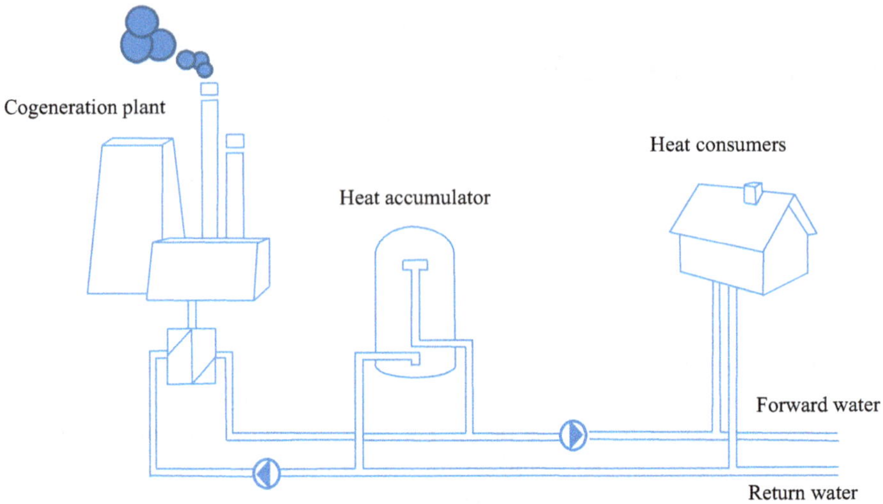

Fig. E.1 Thermal storage directly connected to a district heating system [30]

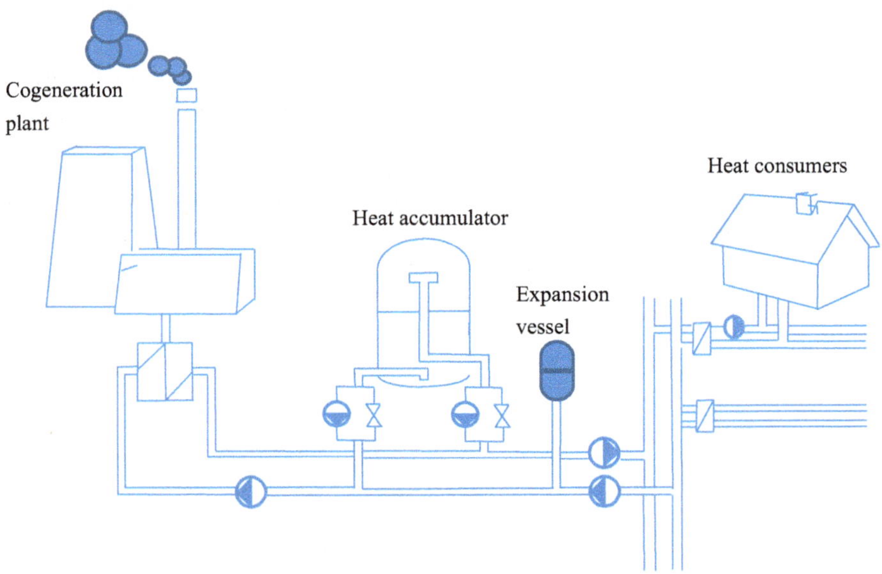

Fig. E.2 Accumulator with hydraulic separation from the district heating system [30]

Appendix F
District Cooling Systems

With the exception of those buildings with a large number of computer equipment and due to climatological reasons, domestic air conditioning, in some parts of Europe, is still being considered a luxury (meanwhile heating is seen as essential).

District cooling systems are similar to district heating networks in the sense that a fluid is distributed and conducted through tubings [31]. However, there are some major distinctions. For example, there are differences between the residential and commercial sectors, which for the latter, are dominated by district cooling networks [32]. This situation can be given the following factors:

- Air conditioning systems are profusely used in the services sector, whereas they have a limited penetration in dwellings [32].
- The size of the district cooling network is a point to be considered so that it is viable from an economic perspective [33].
- Normally, there is a refrigeration density need requirement in order to be profitable to use as a district cooling system, being compulsory that the users are located close to each other [34].
- Another factor is the limited ratio energy/liquid quantity in comparison with DHNs [35] because of the lower temperature differences. For instance, in district cooling systems, flow temperature is around +5 °C, and back temperature is +15 °C (temperature difference of about 10 °C). Meanwhile, in a district heating system, flow temperature would be around 90 °C, and back temperature is around 40 °C (therefore, there is a temperature difference of about 50 °C) [35, 36]. This is the reason why pipes used in district cooling networks are normally higher than those used in district heating networks and thus have a higher cost (Fig. F.1).

Most cooling applications make use of compressors driven by motors [37], where their coefficient of performance depends heavily on the temperature difference [38] (the better the coefficient of performance, the lower the temperature of the heat sink is) [39], as well as the type of compressor, the rating power of the cooling unit, and whether water or air is used to carry out this task [40–43].

A. Colmenar-Santos et al., *District Heating and Cooling Networks in the European Union*, DOI 10.1007/978-3-319-57952-8

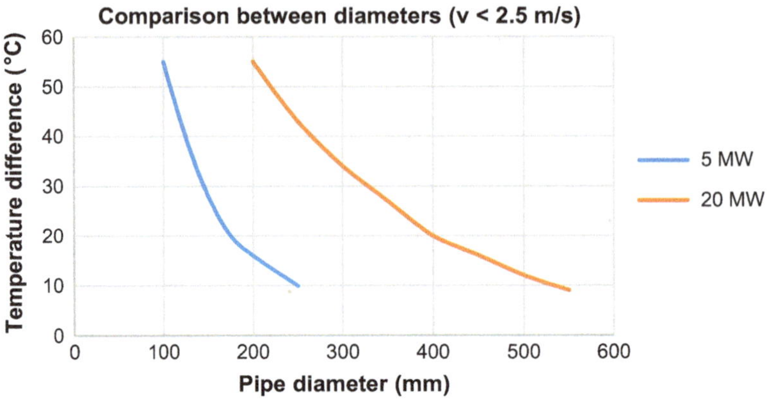

Fig. F.1 Comparison between the pipe diameters taking into account the differences between flow and back temperatures [35]

For its part, the absorption chillers or the heat pumps operate under a different principle [43, 44]. Here, instead of using the mechanical power from a motor, they make use of the heat to drive the cycle [45–48].

It should be noted that the coefficient of performance for this technology lags behind that of compressors driven by electricity [39, 49] and varies for commercially available low-temperature district cooling systems from 0.65 to 1.2 [35]—that is, a unit of heat at low temperature is capable of providing between 0.65 and 1.2 units of low-temperature refrigeration depending primarily on the number of times that the heat entering the chiller is internally used [35]. Traditionally, and due to the elevated temperature's utilisation (those that occur with low-pressure steam) [50], absorption chillers lower the coefficient of performance and the refrigerating capacity greatly as the inlet temperature of the heating medium decreases (Fig. F.2). This has the important effect that for a given cooling capacity, the price of the

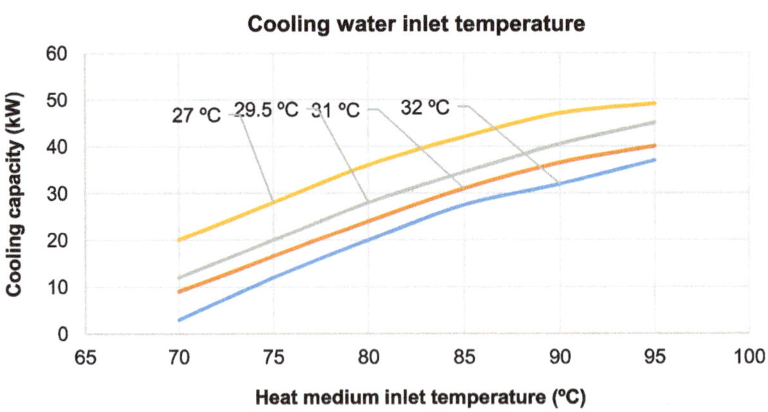

Fig. F.2 Cooling capacity of absorption chillers in relation to water temperature [51]

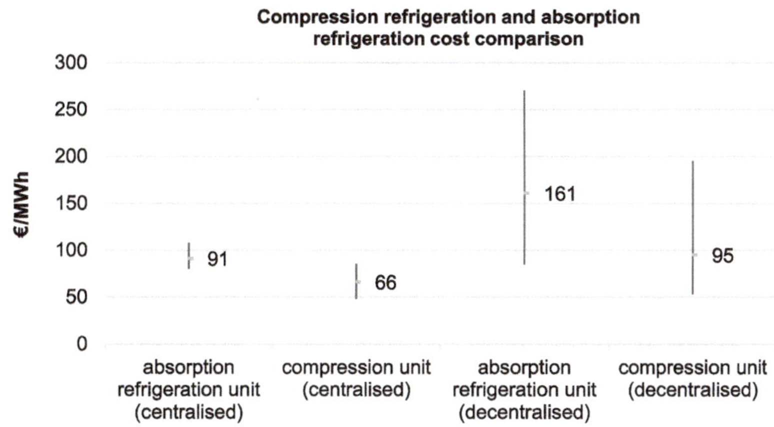

Fig. F.3 Comparison of refrigeration costs through compression and refrigeration through absorption (excluding construction costs of the networks) for power ranging from 300 to 500 kW and 1000 h of operation at full load [52]

device can become excessive (Fig. F.2)—nevertheless, these devices are often in district cooling systems [50].

In (Fig. F.3), it is shown that for 1000 h of operation at full load, cooling costs by absorption (excluding the costs of constructing the networks) are significantly higher than the costs incurred with the technology of refrigeration through compression.

As it is clear from (Fig. F.4) and as far as installation costs are concerned, compression refrigeration is significantly cheaper than refrigeration through absorption. However, as the installed power increases, the differences decrease considerably.

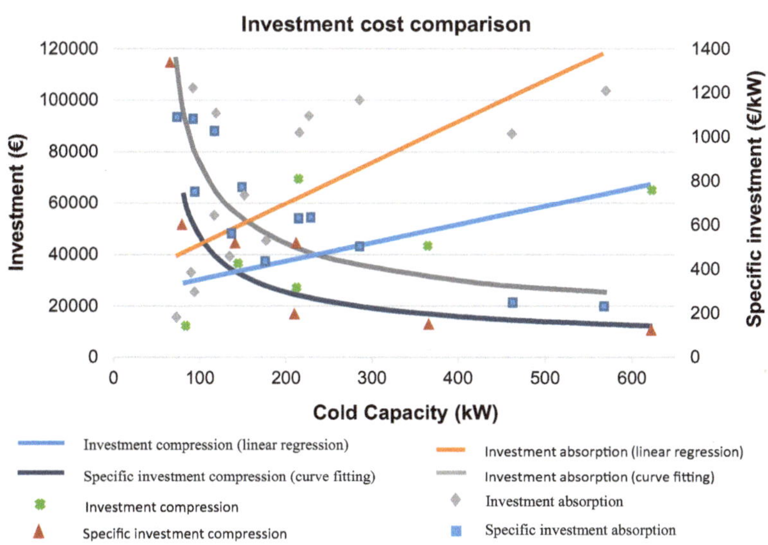

Fig. F.4 Comparison of the investment cost for different refrigeration machines [52]

Appendix G
Fuel Used Invariability and Efficiency

Normally, when using cogeneration power plants and sacrificing a unit of electricity, it is possible to obtain about eight times this energy as heat (in contrast, for example, when using a heat pump, it would only be possible to get that amount multiplied by three) [53–56]. When a plant starts operating as a cogeneration plant and its electricity production falls, another condensation plant which is not operating at maximum output will be responsible for increasing the power in order to compensate for this change, thus resulting in a modification of fuel consumption [57]. By definition, marginal power plants are the ones that dictate the price in every hour to all technologies [58] and, in this case, combustible used to generate electricity will remain constant [59] and at a constant efficiency [60].

A. Colmenar-Santos et al., *District Heating and Cooling Networks in the European Union*, DOI 10.1007/978-3-319-57952-8

Appendix H
Cogeneration, District Heating Networks and Smart Cities: The Case of Copenhagen

The European Commission's "Smart Cities and Communities" initiative has already expressed in the "Strategic Energy Technology Plan" (SET-Plan) an interest in positioning a number of cities at the forefront of combating and minimising CO_2 emissions [61, 62]. District heating and cooling networks will be smarter as they make the electric grid smarter [63, 64]. Consequently, it will be especially important that the heating and cooling have heat reservoirs [65, 66] so that the electric grid reacts quickly to price signals and can stabilise itself when there are fluctuating renewable energy sources.

In Denmark, heating networks are the backbone of the energy system, providing enough flexibility to integrate renewable energies (especially wind) into the electric grid in a safe way [67]. One of the objectives of the EU-28 regarding energy is precisely to make these smart cities adopt the Danish energy strategy [68, 69].

Renewable energy can generate electricity when prices in the electricity market are high, and even though it continues to produce electricity at night when demand is low, the stored heat may cover peak consumption that occurs during the morning. With respect to the loads of heating/cooling, district heating networks and intelligent cooling contribute to smoothing heating and cooling peak demand, thereby reducing peak production [70].

Undoubtedly, Copenhagen is the world region that more has managed to integrate electricity, district heating networks, natural gas, and waste management. Among its highlights may be mentioned the following:

- 98% of the population of Copenhagen is, in those areas where this infrastructure is present, connected to district heating networks [71].
- LAs have developed heating plans for the entire region, dividing the areas by optimal technologies, taking into account economic criteria established at a national level by the Energy Authority [72].
- Municipalities have created companies with a municipal participation (CTR and VEKS) to develop the heat transport system, while companies entirely belonging to the municipality are in charge of distributing this heat.

A. Colmenar-Santos et al., *District Heating and Cooling Networks in the European Union*, DOI 10.1007/978-3-319-57952-8

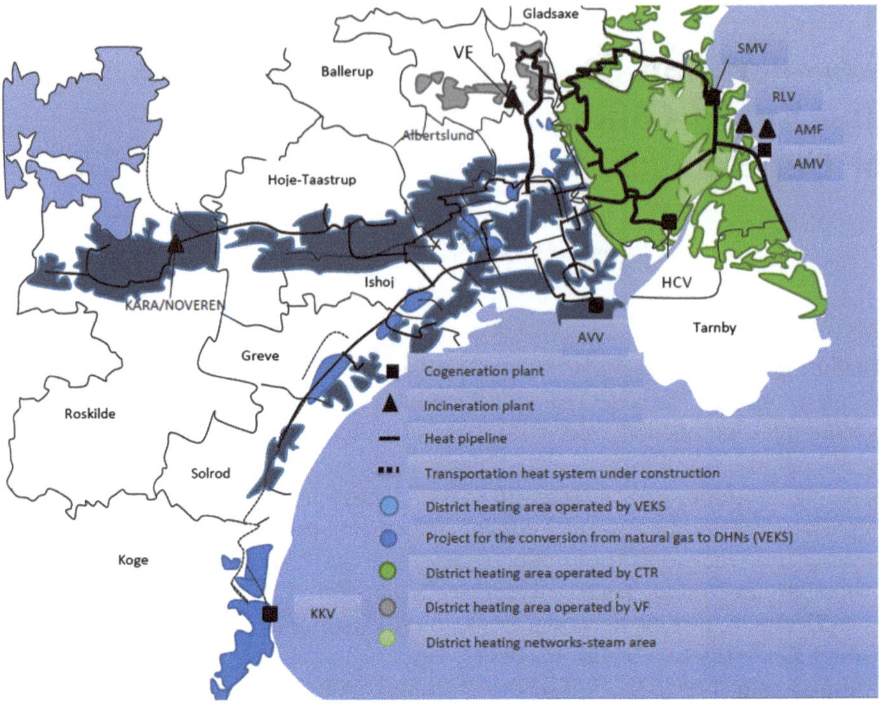

Fig. H.1 Map from VEKS showing the heating system in Copenhagen (only local heating systems in the area of VEKS are presented) [76]

- For its part, the waste management company (Vestforbrænding) is in charge, apart from the recycling, of treating the waste produced by 20 municipalities [7] to an optimum level to introduce cogeneration in large incineration plants to ensure that all the material that cannot be recycled is used as energy.

In Denmark, the providing of the service of district heating depends on companies that are themselves local monopolies, and that typically have the necessary infrastructure for the generation, supply and sale of heat to the end user [73]. In contrast to the companies supplying gas and electricity, for these local monopolies, any benefit from the sale of heat will not be permitted to be obtained [74]. This means that only the necessary costs of construction, operation, and maintenance of district heating networks can finally be included in the fees [75].

To get a visual idea of how the system of generation and heat transport of the city of Copenhagen is, in (Fig. H.1) a map depicting the generation and transmission system operated by VEKS, CTR, and Vestforbraendings is shown.

Appendix I
Extra Information About Extraction-condensing and Backpressure Turbines

While, obviously, gas turbines can be an integral part of cogeneration systems, the truth is that most of them consist of conventional boilers and steam turbines [77]. The two most widely used types of steam turbines are backpressure and extraction-condensing ones, the selection between them being chiefly influenced by economic factors, the amount of electricity to be generated, and the amount (and quality) of heat required.

The simplest configuration is the backpressure turbine. In this case, the pressure at which the mixture of gas and saturated vapour leaves it is equal or superior to the ambient one (this is precisely why the word "backpressure" is used). The most remarkable advantages of the backpressure system are its simple configuration and few components, low capital costs, the fact that it has no need (or, at least, a reduced need) for cooling water, and its high overall efficiency because it does not eject heat outside through a heat condenser [78]. Among the disadvantages of this configuration are that, for the same amount of electricity generated, the steam turbine required is greater (because it operates with steam that has a lower enthalpy difference), and the electricity generated mainly depends on the internal energy requirements (consequently there is a reduced capability to match the electricity generated with the electric load) [78]. However, mainly due to this lack of flexibility, backpressure turbines are typically used in medium-sized cogeneration plants (in fact, the largest backpressure steam turbine built to date has a nominal output of 150 MW) [79], so they do not apply in the analysis conducted in this book, which will focus on large thermal power plants.

By definition, condensing turbines are located exclusively in electricity-only power plants [80]; if they are converted to maintaining the ability of condensing, extracting steam from the low-pressure turbine, and using it in district heating networks, then these turbines are called extraction-condensing turbines [80].

The diagram shown in (Fig. I.1) is used to describe the performance of the plant presented in (Fig. 5.1) by evincing its operational limits and enabling designers to economically evaluate the existing options. When starting in the bottom right corner of the diagram, it is possible to see that the unit at full load can generate 468 MW without producing any amount of useful heat; hence, it emits a large amount of heat into the sea or into the atmosphere.

A. Colmenar-Santos et al., *District Heating and Cooling Networks in the European Union*, DOI 10.1007/978-3-319-57952-8

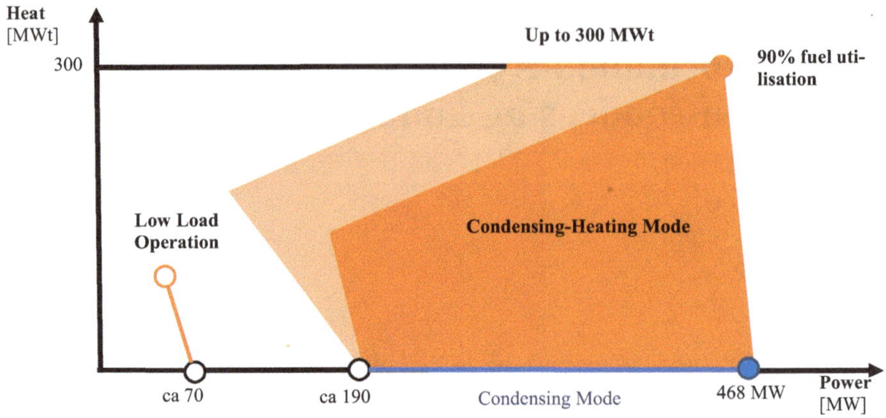

Fig. I.1 Operating limits of an extraction-condensing turbine [81]

As heat from the low-pressure turbine is extracted, the amount of electricity generated decreases, heat begins to be usable, and its point of performance is moved along the right side of the diagram. Finally, when the limit on the top right is reached, about 300 MWt of heat and about 415 MW of electricity are generated.

If the amount of fuel introduced into the boiler is reduced, but it is desired to maximise the amount of heat generated, then the operating point begins to move leftward along the top lines (the turbine would be operating in backpressure mode). In this case, the heat and the electricity would decrease respectively to 90 MWt and 50 MW, being the boiler at this point operating at its minimum amount of fuel used and minimum amount of steam ejected.

If, afterwards, the amount of heat used is reduced, limiting the extraction of steam, then the electricity produced increases, moving the operating point down and to the right until the condensing mode is reached with 190 MW and 0 MWt. If the amount of fuel is augmented enough, the starting point in condensing mode would be reached (468 MW and 0 MWt).

If thermal power is pulled out afresh while the combustible used remains unchanged, then the point of operation would run parallel to the bottom line of the diagram [82–84]. Therefore, from any point on the bottom line of the diagram, it is possible to increase the heat output, decreasing the electricity generated for the same amount of fuel used, and constituting iso-fuel lines (consequently, any operating point within the diagram is feasible) [82–84].

In a system with extraction-condensing turbines, the steam required for internal energy needs is provided by a "bleeding" performed in a set of moving blades behind the nozzle under suitable thermodynamic conditions; this remaining mixture of both gas and saturated vapour is ejected at the pressure of the water-cooled shell

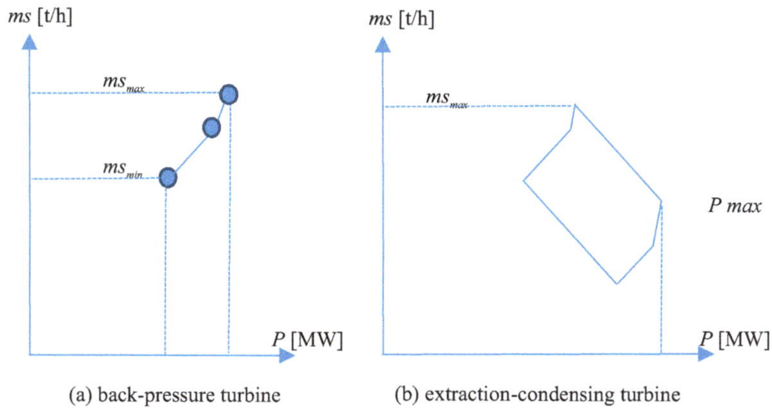

Fig. I.2 Joint production of electricity and heat (mass flow of steam ms measured in t/h). **a** Operation with a backpressure turbine. **b** Flexible operation with an extraction-condensing turbine [85]

and tube heat exchanger [82–84]. Note that although the extraction-condensing turbine type has a higher capital cost and, generally, a lower overall efficiency [79], its capability of modifying output electricity regardless of internal energy requirements (through a specific monitoring and control of the rate of steam) makes them indispensable in large co-generation plants. In (Fig. I.2), the different heat and electricity generation performances in a backpressure turbine and in an extraction-condensing one can be seen.

Appendix J
Comparison of Heat Accumulator Storage for Different District Heating Systems

See Table J.1.

Table J.1 Heat accumulators for different combined heat and power in Danish power plants [86]

City		Odense	Aalborg	Esbjerg
Plant		Fynsværket	Nordjyllandsværket	Vestkraft
Demand for heat to network	GWh/a	2200	2000	1100
Demand for peak load, 3000 h		733	667	367
Demand for base load, 5000 h	MWt	440	400	220
First priority base load waste	MWt	70	80	55
Demand for CHP base load	MWt	370	320	165
Maximum heat capacity—cogeneration plant	MWt	480	460	460
Storage size	m^3	75,000	25,000	55,000
Storage capacity, 45 gr.	MWh	3915	1305	2871
Heat extraction interruption in a winter day—cogeneration plant	hours	11	4	17
Heat extraction interruption in an average day—cogeneration plant	hours	26	11	52
Heat accumulator full load—average time taken in a winter day	hours	36	9	10
Heat accumulator full load—average time taken in an average day	hours	12	4	7

A. Colmenar-Santos et al., *District Heating and Cooling Networks in the European Union*, DOI 10.1007/978-3-319-57952-8

Appendix K
Supporting Data for "Final Results" Tables (Tables 5.9–5.14)

See Tables K.1, K.2, K.3, K.4, K.5, K.6, K.7, K.8, K.9, K.10, K.11.

A. Colmenar-Santos et al., *District Heating and Cooling Networks in the European Union*, DOI 10.1007/978-3-319-57952-8

Table K.1 Annualised costs for the base case of Oldenburg-Wilhelmshaven at a discount rate of 3.5%

				Annualised costs				
	Installed power	Number of units	Annual costs	Cost of capital	Operational and maintenance costs—fixed	Operational and maintenance costs—variable	Fuel costs	Revenues—electricity
	MW	units	k€	k€	k€	k€	k€	k€
Existing electrical power plant	757	1	−9635	0	54,504	10,941[a]	152,160	227,240
Individual gas boiler (dwellings)	829	92,080	119,255	3670	2818	0	112,767	0
Individual gas boiler (other cons.)	590	22,720	89,687	2612	2006	0	85,069	0
Individual electric heating (dwellings)	207	23,020	133,690	3954	4347	0	125,389	0
Individual electric heating (other cons.)	150	5680	99,713	2865	3150	0	93,698	0
Individual air conditioning	0	0	0	0	0	0	0	0
	Total annual costs		432,710					

Source Own elaboration
[a]A variable cost of operation and maintenance of 2.5 €/MWh has been assumed [87]

Table K.2 Annualised costs for the Case COGEN-a1 of Oldenburg-Wilhelmshaven at a discount rate of 3.5%

				Annualised costs				
	Installed power	Number of units	Annual costs	Cost of capital	Operational and maintenance costs—fixed	Operational and maintenance costs—variable	Fuel costs	Revenues—electricity
	MW	Units	k€	k€	k€	k€	k€	k€
Cogeneration plant conversión	757	1	32,945	12,138	54,504	10,941	152,160	196,798
Transmission pipeline (heat)	1789		38,675	30,982	7693	0	0	0
Heat accumulator		1	349	349	0	0	0	0
District heating network (dwellings)	1036	115,100	52,474	24,917	27,557	0	0	0
District heating network (other cons.)	753	28,400	26,177	6148	20,029	0	0	0
Wet system (dwellings)	207	23,020	2109	2109	0	0	0	0
Wet system (other cons.)	150	5680	520	520	0	0	0	0
Centralised gas boiler	931		33,127	4302	2095	1396	25,334	0
Individual heat exchanger (dwellings)	1036	115,100	47,876	41,660	6216	0	0	0
Individual heat exchanger (other cons.)	753	28,400	14,797	10,279	4518	0	0	0
Individual air conditioning	0	0	0	0	0	0	0	0
	Total annual costs		249,049					

Source Own elaboration

Table K.3 Annualised costs for the Case COGEN-a2 of Oldenburg-Wilhelmshaven at a discount rate of 3.5%

				Annualised costs				
	Installed power	Number of units	Annual costs	Cost of capital	Operational and maintenance costs—fixed	Operational and maintenance costs—variable	Fuel costs	Revenues—electricity
	MW	units	k€	k€	k€	k€	k€	k€
New cogeneration plant	757		75,432	54,625	54,504	10,941	152,160	196,798
Transmission pipeline (heat)	1789		38,675	30,982	7693	0	0	0
Heat accumulator		1	349	349	0	0	0	0
District heating network (dwellings)	1036	115,100	52,474	24,917	27,557	0	0	0
District heating network (other cons.)	753	28,400	26,177	6148	20,029	0	0	0
Wet system (dwellings)	207	23,020	2109	2109	0	0	0	0
Wet system (other cons.)	150	5680	520	520	0	0	0	0
Centralised gas boiler	931		33,127	4302	2095	1396	25,334	0
Individual heat exchanger (dwellings)	1036	115,100	47,876	41,660	6216	0	0	0
Individual heat exchanger (other cons.)	753	28,400	15,884	10,279	4518	0	0	0
Individual air conditioning	0	0	0	0	0	0	0	0
	Total annual costs		292,623					

Source Own elaboration

Table K.4 Annualised costs for the Base Case of Bristol at a discount rate of 3.5%

				Annualised costs				
	Installed power	Number of units	Annual costs	Cost of capital	Operational and maintenance costs—fixed	Operational and maintenance costs—variable	Fuel costs	Revenues—electricity
	MW	units	k€	k€	k€	k€	k€	k€
Existing electrical power plant	1140 (755 +385)	2	−53,020	0	35,340	13,400[a]	280,140	381,900
Individual gas boiler (dwellings)	781	86,800	95,249	3458	2655	0	89,136	0
Individual gas boiler (other cons.)	575	21,680	71,656	2456	1955	0	67,245	0
Individual electric heating[b] (dwellings)	195	21,700	78,759	3725	4095	0	70,939	0
Individual electric heating (other cons.)	144	5420	59,291	2751	3024	0	53,516	0
Individual air conditioning	0	0	0	0	0	0	0	0
	Total annual costs		251,935					

Source Own elaboration

[a]A variable cost of operation and maintenance of 2 €/MWh has been assumed [88]

[b]For the Base Case, it has been assumed that 20% of users use individual electric heating [90]

Table K.5 Annualised costs for the Case COGEN-a1 of Bristol at a discount rate of 3.5%

				Annualised costs				
	Installed power	Number of units	Annual costs	Cost of capital	Operational and maintenance costs—fixed	Operational and maintenance costs—variable	Fuel costs	Revenues—electricity
	MW	units	k€	k€	k€	k€	k€	k€
Cogeneration plant conversión	1140 (755+385)	2	−5840	9140	35,340	13,400	280,140	343,860
Transmission pipeline (heat)	1695		1258	9785	2373	0	0	0
Heat accumulator		1	349	349	0	0	0	0
District heating network (dwellings)	977	108,500	49,476	23,488	25,988	0	0	0
District heating network (other cons.)	718	27,100	24,965	5867	19,098	0	0	0
Wet system (dwellings)	195	21,700	1988	1988	0	0	0	0
Wet system (other cons.)	144	5420	497	497	0	0	0	0
Centralised gas boiler	878		31,821	4057	1989	1303	24,472	0
Individual heat exchanger (dwellings)	977	108,500	45,133	39,271	5862	0	0	0
Individual heat exchanger (other cons.)	718	27,100	14,116	9808	4308	0	0	0
Individual air conditioning	0	0	0	0	0	0	0	0
	Total annual costs		174,663					

Source Own elaboration

Table K.6 Annualised costs for the Case COGEN-a2 of Bristol at a discount rate of 3.5%

				Annualised costs				
	Installed power	Number of units	Annual costs	Cost of capital	Operational and maintenance costs—fixed	Operational and maintenance costs—variable	Fuel costs	Revenues—electricity
	MW	units	k€	k€	k€	k€	k€	k€
New cogeneration plant	1140		20,438	35,418	35,340	13,400	280,140	343,860
Transmission pipeline (heat)	1695		12,158	9785	2373	0	0	0
Heat accumulator		1	349	349	0	0	0	0
District heating network (dwellings)	977	108,500	49,476	23,488	25,988	0	0	0
District heating network (other cons.)	718	27,100	24,965	5867	19,098	0	0	0
Wet system (dwellings)	195	21,700	1988	1988	0	0	0	0
Wet system (other cons.)	144	5420	497	497	0	0	0	0
Centralised gas boiler	878		31,821	4057	1989	1303	24,472	0
Individual heat exchanger (dwellings)	977	108,500	45,133	39,271	5862	0	0	0
Individual heat exchanger (other cons.)	718	27,100	14,116	9808	4308	0	0	0
Individual air conditioning	0	0	0	0	0	0	0	0
	Total annual costs		209,941					

Source Own elaboration

Table K.7 Annualised costs for the Base Case of Cartagena at a discount rate of 3.5%

				Annualised costs				
	Installed power	Number of units	Annual costs	Cost of capital	Operational and maintenance costs—fixed	Operational and maintenance costs—variable	Fuel costs	Revenues—electricity
	MW	Units	k€	k€	k€	k€	k€	k€
Existing electrical power plant	1200 (3*400)	3	−300	0	37,200	14,100	307,440	359,040
Individual gas boiler (dwellings)	598	92,000	59,371	2648	2033	0	54,690	0
Individual gas boiler (other cons.)	441	23,200	44,708	1952	1499	0	41,257	0
Individual electric heating (dwellings)	150	23,000	42,207	2865	3150	0	36,192	0
Individual electric heating (other cons.)	110	5800	31,714	2101	2310	0	27,303	0
Individual air conditioning (dwellings)[a]	190	54,050	139,329	9205	3496	0	126,628	0
Individual air conditioning (other cons.)	100	13,630	73,331	4845	1840	0	66,646	0
	Total annual costs		390,360					

Source Own elaboration

[a]It was considered that, for the city of Cartagena, the penetration of air conditioning is 47% [91] with a rated power of 3.2 kW for dwellings and 7 kW for other consumers [92]

Table K.8 Annualised costs for the Case COGEN-a1 of Cartagena at a discount rate of 3.5%

				Annualised costs				
	Installed power	Number of units	Annual costs	Cost of capital	Operational and maintenance costs—fixed	Operational and maintenance costs—variable	Fuel costs	Revenues—electricity
	MW	units	k€	k€	k€	k€	k€	k€
Cogeneration plant conversión	1200	3	38,630	9621	37,200	14,100[a]	307,440	328,804
Transmission pipeline (heat)	1299		9317	7499	1818	0	0	0
Heat accumulator		1	349	349	0	0	0	0
District heating network (dwellings)	748	115,000	44,791	24,895	19,896	0	0	0
District heating network (other cons.)	551	29,000	20,934	6278	14,656	0	0	0
Wet system (dwellings)	150	23,000	2107	2107	0	0	0	0
Wet system (other cons.)	110	5800	531	531	0	0	0	0
Centralised gas boiler	396		8067	1830	891	594	4752	0
Individual heat exchanger (dwellings)	748	115,000	46,112	41,624	4488	0	0	0
Individual heat exchanger (other cons.)	551	29,000	13,803	10,497	3306	0	0	0
Individual air conditioning	190	54,050	139,329	9205	3496	0	126,628	0
	100	13,630	73,331	4845	1840	0	66,646	0
	Total annual costs		397,301					

Source Own elaboration
[a]A variable cost of operation and maintenance of 2 €/MWh has been assumed [89]

Table K.9 Annualised costs for the Case COGEN-a2 of Cartagena at a discount rate of 3.5%

				Annualised costs				
	Installed power	Number of units	Annual costs	Cost of capital	Operational and maintenance costs—fixed	Operational and maintenance costs—variable	Fuel costs	Revenues—electricity
	MW	units	k€	k€	k€	k€	k€	k€
New cogeneration plant	1200		67,218	37,282	37,200	14,100	307,440	328,804
Transmission pipeline (heat)	1299		9317	7499	1818	0	0	0
Heat accumulator		1	349	349	0	0	0	0
District heating network (dwellings)	748	115,000	44,791	24,895	19,896	0	0	0
District heating network (other cons.)	551	29,000	20,934	6278	14,656	0	0	0
Wet system (dwellings)	150	23,000	2107	2107	0	0	0	0
Wet system (other cons.)	110	5800	531	531	0	0	0	0
Centralised gas boiler	396		8067	1830	891	594	4752	0
Individual heat exchanger (dwellings)	748	115,000	46,112	41,624	4488	0	0	0
Individual heat exchanger (other cons.)	551	29,000	13,803	10,497	3306	0	0	0
Individual air conditioning (dwellings)	190	54,050	139,329	9205	3496	0	126,628	0
Individual air conditioning (other cons.)	100	13,630	73,331	4845	1840	0	66,646	0
	Total annual costs		425,889					

Source Own elaboration

Table K.10 Annualised costs for the Case COGEN-b of Cartagena at a discount rate of 3.5%

				Annualised costs				
	Installed power	Number of units	Annual costs	Cost of capital	Operational and maintenance costs—fixed	Operational and maintenance costs—variable	Fuel costs	Revenues—electricity
	MW	units	k€	k€	k€	k€	k€	k€
Cogeneration plant conversión	1200	3	38,630	9621	37,200	14,100[a]	307,440	328,804
Transmission pipeline (heat)	1589		11,398	9173	2225	0	0	0
Heat accumulator		1	349	349	0	0	0	0
District heating network (dwellings)	921	115,000	48,994	24,495	24,499	0	0	0
District heating network (other cons.)	646	29,000	22,855	5672	17,183	0	0	0
Wet system 580	150	23,000	1404	1404	0	0	0	0
Wet system (other cons.)	110	5800	354	354	0	0	0	0
Centralised gas boiler	396		11,451	1830	891	594	8136	0
Individual absorption unit (dwelling)	190	115,000	8505	8182	323	0	0	0
Individual absorption unit (other cons.)	100	29,000	4477	4307	170	0	0	0
Individual heat exchanger (dwellings)	748	115,000	46,112	41,624	4488	0	0	0
Individual heat exchanger (other cons.)	551	29,000	13,803	10,497	3306	0	0	0
	Total annual costs		208,332					

Source Own elaboration

[a]A variable cost of operation and maintenance of 2 €/MWh has been assumed [89]

Table K.11 Annualised costs for the Case COGEN-c of Cartagena at a discount rate of 3.5%

				Annualised costs				
	Installed power	Number of units	Annual costs	Cost of capital	Operational and maintenance costs—fixed	Operational and maintenance costs—variable	Fuel costs	Revenues—electricity
	MW	units	k€	k€	k€	k€	k€	k€
Cogeneration plant conversión	1200	3	38,630	9621	37,200	14,100	307,440	328,804
Transmission pipeline	1299		9317	7499	1818	0	0	0
Accumulator		1	349	349	0	0	0	0
District heating network (dwellings)	748	115,000	44,391	24,495	19,896	0	0	0
District heating network (other cons.)	551	29,000	20,328	5672	14,656	0	0	0
District cooling network (dwellings)	190	115,000	29,949	24,895	5054	0	0	0
District cooling network (other cons.)	100	29,000	8938	6278	2660	0	0	0
Wet system (dwellings)	150	23,000	2107	2107	0	0	0	0
Wet system (other cons.)	110	5800	531	531	0	0	0	0
Centralised gas boiler	396		11,451	1830	891	594	8136	0
Centralised absorption unit	369		17,564	1183	1180	4331	10,870[a]	0
Individual heat exchanger (dwellings)	748	115,000	46,112	41,624	4488	0	0	0

(continued)

Table K.11 (continued)

				Annualised costs				
	Installed power	Number of units	Annual costs	Cost of capital	Operational and maintenance costs—fixed	Operational and maintenance costs—variable	Fuel costs	Revenues—electricity
	MW	units	k€	k€	k€	k€	k€	k€
Individual heat exchanger (other cons.)	551	29,000	13,803	10,497	3306	0	0	0
Individual fan coil (dwellings)	190	115,000	14,126	826	13,300	0	0	0
Individual fan coil (other cons.)	100	29,000	7435	435	7000	0	0	0
	Total annual costs		265,031					

Source Own elaboration
[a]See [93]

References

1. IEA. World balance. Paris: IEA. http://www.iea.org/sankey/. Accessed 22 Sept 2015.
2. Euroheat & amp; Power. District heating and cooling country by country survey 2015. Brussels: Euroheat & Power; 2015.
3. Orchard W. Cost of transmitting 2 GW of heat 140 km: Also longest heat main in Europe 60 km. London: Orchard Partners London Ltd. 2011. https://groups.google.com/forum/#!msg/energy-discussion-group/hOqhtmoHic8/0BrOuaIJ3V8J. Accessed 11 Mar 2014.
4. Pirouti M. Modelling and analysis of a district heating network. Cardiff: University of Cardiff; 2013. p. 137.
5. Kitsios K, Chrysis I. Assessment of national potential for cogeneration in Cyprus. Nicosia: Republic of Cyprus Ministry of Commerce, Industry and Tourism Energy Service; 2009.
6. Riddle A. District heating & the future smart cities approach. Copenhagen: Ramboll Group A/S; 2013.
7. Dyrelund A. The Future of the energy supply: Smart energy cities. Euroheat & Power 2012 (English Edition);9(1):12–15. https://www.stateofgreen.com/CMSPages/GetAzureFile.aspx?path= ~ %5Ccache%5Cstateofgreen%5C33%5C3305a159-39c0-4c5f-ae58-70883774a7d4.pdf&hash=3e2d92789d181deb3076da90e24ce0abe6548c73131c70d1d8e172d7661ebec9. Accessed 11 Mar 2014.
8. Worm J. Delrapport 3 miscellaneous investigations: CO_2 reductions in low-energy buildings and communities by implementation of low-temperature district systems—demonstration cases in Energyflexhouse and Boligforeningen Ringgården, Journalnr. 63011-0152. Copenhagen: Danish Energy Agency. 2011. http://fjvu.dk/sites/default/files/delrapport_3_miscellaneous_investigations.pdf. Accessed 11 Mar 2014.
9. Energy Dong. The Studstrupværket CHP plant. Skødstrup: Dong Energy A/S; 2007.
10. Wagner I. Local success stories: Prague district heating system. Brussels: Ecoheat4.eu. 2011. http://ecoheat4.eu/en/Country-by-country-db/Czech-Republic/Local-Success-Stories/. Accessed 11 Mar 2014.
11. Piskac M. Prague district heating system: application for 1st Global District Energy Climate Awards. Praga: Pražská Teplárenská A.S. 2009. http://www.copenhagenenergysummit.org/applications/Prague%20Czech%20Rep-District%20Energy%20Climate%20Award.pdf. Accessed 11 Mar 2014.
12. Harvey LDD. Energy and the new reality 1: Energy efficiency and the demand for energy services. 1ª ed. Oxon: Routledge; 2010.
13. Wade DW. District heating conversion from steam to hot water at the Savannah Regional Hospital. Marietta: RDA Engineering; 1995.
14. De Souza GFM. Thermal power plant performance analysis. 1st ed. London: Springer-Verlag, London Limited; 2012.

15. Burzynski R, Crane M, Yao R, Becerra VM. Low temperature district heating network serving experimental zero carbon homes in Slough, UK. Copenhagen: DHC13, the 13th International Symposium on District Heating and Cooling; 2012.
16. Energy Ramboll. Lavenergi Fjernvarmekoncepter. Copenhagen: Ramboll; 2010.
17. Urecon. Pre-insulated municipal service pipe system. Quebec: Urecon Ltd. 2009. http://www.urecon.com/applications/municipal_pipe.html. Accessed 17 Mar 2014.
18. Eliseev K. District heating systems in Finland and Russia. Mikkeli: Mikkeli University of Applied Sciences; 2011.
19. Hepbasli A. Low exergy (LowEx) heating and cooling systems for sustainable buildings and societies. Renew Sustain Energ Rev. 2012;16:74–104.
20. Allegrini J, Orehounig K, Mavromatidis G, Ruesch F. A review of modelling approaches and tools for the simulation of district-scale energy systems. Renew Sustain Energ Rev. 2015;52:1391–404.
21. Pomper DE. Electric storage: Technologies and regulation. Silver Spring: National Regulatory Research Institute; 2011.
22. Pauli GA. The blue economy: 10 Years, 100 innovations, 100 million jobs. 1st ed. Taos: Paradigm Publications; 2010.
23. Energinet. Technology data for energy plants: Generation of electricity and district heating, energy storage and energy carrier generation and conversion. Erritsø: Energinet; 2012.
24. Taylor P, Bolton R, Stone D, Zhang XP, Martin C, Upham P. Pathways for energy storage in the UK. York: The Centre for Low Carbon Futures; 2012.
25. Directorate-General for Energy. The future role and challenges of energy storage. Brussels: European Commission; 2013.
26. Macnaghten J. Utility scale pumped heat electricity storage. Fareham: Isentropic Ltd; 2009.
27. IEA-ETSAP. Thermal energy storage: Technology brief. Bonn: International Renewable Energy Agency; 2013.
28. Schröder D. Introducing additional heat storage to the hässelby chp plant: A case study on economic and ecological benefits achievable with heat storage in a deregulated electricity market. Stockholm: KTH School of Industrial Engineering and Management; 2011.
29. International Energy Agency. Combined heat and power: Evaluating the benefits of greater global investment. París: International Energy Agency; 2008.
30. ESMAP. Increasing the energy efficiency of heating systems in Central and Eastern Europe and the former Soviet Union. Washington: The World Bank; 2000.
31. Poeuf P, Senejean B, Ladaurade C. District cooling system: the most efficient system for urban applications. París: Climespace—GDF Suez; 2010.
32. DHC+ Technology Platform. District heating & cooling: a vision towards 2020–2030–2050. Brussels: DHC+ Technology Platform; 2009.

33. Thornton R. IDEA report: the district energy industry. Westborough: International District Energy Association; 2005.
34. Cooling Capital. EU district cooling market and trends. Stockholm: Capital Cooling; 2013.
35. Skagestad B, Mildenstein P. District heating and cooling connection handbook. Paris: International Energy Agency. IEA District Heating and Cooling; 2001.
36. Boysen H. Hydraulic balance in a district cooling system. Nordborg: Danfoss district energy; 2003.
37. Wulfinghoff DR. Energy efficiency manual. 1st ed. Montgomery County: Energy Institute Press; 1999.
38. Kavanaugh S. How air-conditioners and heat pumps work. Northport: Energy Information Services; 2011.
39. Medved S. Energy production: Generators of cold chillers. Maastrich: Cauberg-Huygen Consulting Engineers BV; 2013.
40. Jamadar VM, Patil AM. Investigation of performance analysis of compressor cooling in vapour compression system. Int J Adv Eng Res Stud. 2012;2 (1):29–31.
41. Sellars B, Barbour A. Technical report: geothermal heat pump study. Auckland: Beca Carter Hollings & Ferner Ltd.; 2009.
42. Tovey K. ENV-2D02 5: Section 11. Norwich: University of East Anglia; 2006.
43. Baechler MC. Energy renovations HVAC: a guide for contractors to share with homeowners. Richland: Pacific Nothwest National Laboratory; 2011.
44. Zimny J, Michalak P, Szczotka K. Polish heat pump market between 2000 and 2013: European background, current state and development prospects. Renew Sustain Energ Rev. 2015;48:791–812.
45. Jolly A. Energy efficiency of buildings and renewable energy. Hurup: Nordic Folkecenter for Renewable Energy; 2006.
46. U.S. Department of Energy. Use low-grade waste steam to power absorption chillers. Washington: Industrial Technologies Program Energy Efficiency; 2006.
47. Srikhirin P, Aphornratana S, Chungpaibulpatana S. A review of absorption refrigeration technologies. Renew Sustain Energ Rev. 2001;5(4):343–72 Amsterdam.
48. Energy Management Association of New Zealand. Energy audit standard for process heat systems: a standard for the auditing of the energy efficiency of direct and indirect heating systems. Wellington: Energy Management Association of New Zealand; 2012.
49. Brown RBY. An experimental study of heat driven absorption cooling systems. Salford: University of Salford; 1990.
50. Labus J. Modelling of small capacity absorption chillers driven by solar thermal energy or waste heat. Tarragona: Universitat Rovira i Virgili; 2011.
51. ATECYR. Equipos de absorción. Madrid: Asociación Española de Climatización y Refrigeración; 2008.
52. Hübner V, Kallmann K, Piller S, Thamling N, Hesse D, Wetzel A. Meeting cooling demands in summer by applying heat from cogeneration. Berlín:

Berliner Energieagentur GmbH. 2008. http://www.chaleurfraicheur.org/doc/Publishable_Report_FINAL.pdf. Accessed 8 Mar 2014.
53. Basel II, Ahmed WH. Thermoelectric power generation using waste-heat energy as an alternative green technology. Recent Pat Electr Eng. 2009;2:27–39.
54. ICF International. Technology characterization: steam turbines. Washington DC: Environ Prot Agency; 2008.
55. Elsaket G. Simulating the integrated solar combined cycle for power plants application in Libya. Cranfield: Cranfield University; 2007.
56. Lowe R. Combined heat and power considered as a virtual steam cycle heat pump. Energ Policy. 2011;39:5528–34.
57. Fahlen E, Ahlgren EO. Accounting for external costs in a study of a Swedish district-heating system: an assessment of environmental policies. Energ Policy. 2010;28:4909–20.
58. Cortes Generales. Diario de sesiones del Congreso de los Diputados, 8 de marzo de 2006. Núm. 507. Madrid: Congreso de los Diputados; 2006.
59. UNEP. Thermal energy equipment. www.retscreen.net/fichier.php/1000. Accessed 17 Dec 2013.
60. Uche-Marcuello FJ. Thermoeconomic analysis and simulation of a combined power and desalination plant. Zaragoza: Universidad de Zaragoza; 2000.
61. Climate and Energy Fund. Smart cities and initiative. Vienna: Climate and Energy Fund; 2013.
62. European technology platform on renewable heating and cooling. Strategic research and innovation agenda for renewable heating & cooling. Luxembourg: European Union; 2013.
63. Gadd H. To measure is to know! Lund: Lund University; 2012.
64. DHC+. District heating & cooling: strategic research agenda. Brussels: Euroheat & Power; 2012.
65. Møller-Jørgensen J. R&D storage activities in Denmark and EU. Fredericia: Energinet; 2013.
66. Wu W, Wang B, Shi W, Li X. An overview of ammonia-based absorption chillers and heat pumps. Renew Sustain Energ Rev. 2014;31:681–707.
67. Kristensen S. The backbone of the modern energy system. Frederiksberg: Danish Board of District Heating. 2010. http://dbdh.dk/the-backboneof-the-modern-energy-system/. Accessed 17 Jan 2014.
68. Dyrelund A. Smart energy cities: Seminar in DG17 28.09.2011. https://stateofgreen.com/files/download/325. Accessed 17 Dec 2013.
69. DESIRE-project. Guidelines to promote chp concepts with heat accumulators and the perspective of CHP plants and other technologies that use thermal energy storage and their implementation in the European Union. Aalborg: Aalborg University; 2007.
70. Schmidt R, Fevrier N, Dumas P. Key to innovation integrated solution: smart thermal grids. Brussels: Smart Cities Stakeholder Platform; 2013.
71. Elsman P. Conpenhagen district heating system. Copenhagen: First Global District Energy Climate Awards; 2009.

72. Elleriis J. Heat planning in the greater Copenhagen area. Copenhagen: Heat Plan Copenhagen; 2011.
73. Dyrelund A, Kozùowski W. DH/CHP in EU smart energy cities. Wrocław: Ramboll; 2012.
74. Danish Ministry of Energy, Utilities and Climate. Electricity, district heating and natural gas supply. Copenhagen: Danish Ministry of Energy, Utilities and Climate. 2012. http://www.efkm.dk/en/climate-energy-and-building-policy/denmark/energy-supply-and-efficiency/electricity-district-heating. Accessed 30 Sept 2015.
75. DBDH. Forsyning Helsingør: Portfolio items. http://dbdh.dk/portfolio-items/forsyning-helsingor/. Accessed 30 Sept 2015.
76. VEKS. Annual report 2012. Albertslund: VEKS; 2013.
77. U.S. Department of Energy. Best practices steam end user training guide: alternate text narratives and graphic descriptions. Washington: Office of Energy Efficiency & Renewable Energy; 2010.
78. UNEP. Cogeneration: Presentation from the "Energy efficiency guide for industry in Asia"—training session on energy equipment. Nairobi: United Nations Environment Programme; 2006.
79. Leyzerovich AS. Steam turbines for modern fossil-fuel power plants. Boca Ratón: The Fairmont Press; 2008.
80. Sørensen RM. Parallel session: combined heat and power—Task 9.3 dissemination report; 2013.
81. Alstom. KA26 ecoHEAT TM combined heat and power production with a combined-cycle power plant. Baden: Alstom; 2012.
82. Gargiulo M. Getting started with TIMES-VEDA: Version 2,7. Paris: IEA—Energy Technology Systems Analysis Programme; 2009.
83. Capros P, Mantzos L, Criqui P, Kouvaritakis N, Soria-Ramírez A, Schrattenholzer L, et al. Climate technology strategies 1: controlling greenhouse gases—policy and technology options. Heidelberg: Physica-Verlag; 1999.
84. Cosijns L, Haeseldonckx D, D'haeseleer WD, Grohnheit PE. Eusustel: work package 3 electricity generation technologies and system integration—Subtask 3.1.3 combined heat & power. Leuven: Katholieke Universiteit Leuven; 2006.
85. Baumgärtner S. Ambivalent joint production and the natural environment: an economic and thermodynamic analysis. Heidelberg: Physica-Verlag; 2000.
86. Dyrelund A. Heat/thermal storage in accumulators in DH schemes. http://xa.yimg.com/kq/groups/20593576/1120597166/name/heat. Accessed 29 May 2015.
87. Rogers M, Duffy A. Engineering project appraisal. 2ª ed. Hoboken: Wiley-Blackwell; 2012.
88. Laitner JA. Energy efficiency investments as an economic productivity strategy for Texas. Washington: American Council for an Energy-Efficient Economy; 2011.

89. Kehlhofer R, Hannemann F, Stirnimann F, Rukes B. Combined-cycle gas & steam turbine power plants. 3rd ed. Tulsa: PennWell Corporation; 2009.
90. Kemna R, Van Elburg M, Li W, Van Holsteijn R. Eco-design of boilers: task 3 report (Final)—consumer behaviour & local infrastructure. Delft: Van Holsteijn en Kemna BV; 2007.
91. Holley AM. Global trends in air conditioning. Bracknell: BSRIA Ltd.; 2014.
92. Emerson Network Power. Data center precision cooling: the need for a higher level of service expertise. Westerville: Emerson Network Power; 2010.
93. Matsushuta S, Fukushima R, Fujiwara M, Otake M. High-performance absorption chiller for district heating and cooling. Mitsubishi Heavy Ind Tech Rev. 2002;39(2):73–8.

Bibliography

1. Greenwood R, Raynard M, Kodeih F, Micelotta ER, Lounsbury F. Institutional complexity and organizational responses. The Acad Manag Ann. 2011;5(1):317–71.
2. Dewick P, Foster C. Focal actors and ecoinnovation. Copenhagen: DRUID Society Conference 2011; 2011.
3. Sederlöf J. The marketing communications industry in flux: large B2C advertisers' viewpoint. Turku: Turku School of Economics; 2012.
4. Wojewnik-Filipkowska A. Public private cooperation in sustainable city development—the case study of public-private partnership in railway station area regeneration project. Rome: knowing to manage the territory, protect the environment, evaluate the cultural heritage, FIG working week 2012; 2012.
5. CEER. Status review of renewable and energy efficiency support schemes in Europe. Brussels: Council of European Energy Regulators; 2013.
6. Kontogianni A, Tourkolias CH, Damigos D, Skourtos M. Assessing sea level rise costs and adaptation benefits under uncertainty in Greece. Environ Sci Policy. 2014;37:61–78.
7. Bhatia A. Overview of vapor absorption cooling systems. Stony Point: continuing Education and Development; 2011.
8. Dyrelund A. European energy efficiency legislation adopts 35 years of Danish experience. Copenhagen: Ramboll; 2011.
9. Ministry of Environment and Energy. Denmark's second national communication on climate change. Copenhagen: Danish Environmental Protection Agency; 1997.

A. Colmenar-Santos et al., *District Heating and Cooling Networks in the European Union*, DOI 10.1007/978-3-319-57952-8

FSC
www.fsc.org
MIX
Papier | Fördert
gute Waldnutzung
FSC® C083411

Kolibri 360 GmbH
Ferdinand-Jühlke-Straße 7
99095 Erfurt, Deutschland
produktsicherheit@kolibri360.de